Worlds on Fire
Volcanoes on the Earth, the Moon, Mars, Venus and Io

Worlds on Fire takes the reader on a fascinating tour of the
mightiest volcanoes in the Solar System. From Kilauea volcano in
Hawaii and Mount Etna in Sicily, it leaps to the lava fields and rilles
of the Moon, retraces the historic footsteps of the *Apollo* astronauts
and describes new volcanic provinces to explore. The three largest
volcanoes of Mars – Olympus Mons, Alba Patera and Arsia Mons –
are profiled, amongst others. The strange world of Venus, revealed
by radar, opens our perspective of volcanism to features never seen
before: pancake domes of puffed-up lava, and gigantic fault rings
sitting over buried magma chambers. The tour of the Solar System
ends with the only current eruptions outside of Earth: the
spectacular volcanoes of Io – Jupiter's fiery moon. This highly
readable book, illustrated with the most recent imagery from
spacecraft, will appeal to general readers, and students of Earth and
planetary sciences.

CHARLES FRANKEL is a geologist who has written several books
and directed scientific documentaries.

Worlds on Fire

Volcanoes on the Earth, the Moon, Mars, Venus and Io

CHARLES FRANKEL

CAMBRIDGE
UNIVERSITY PRESS

CAMBRIDGE UNIVERSITY PRESS
Cambridge, New York, Melbourne, Madrid, Cape Town, Singapore, São Paulo, Delhi

Cambridge University Press
The Edinburgh Building, Cambridge CB2 8RU, UK

Published in the United States of America by Cambridge University Press, New York

www.cambridge.org
Information on this title: www.cambridge.org/9780521803939

First published 2005
This digitally printed version (with corrections) 2008

A catalogue record for this publication is available from the British Library

ISBN 978-0-521-80393-9 hardback
ISBN 978-0-521-00863-1 paperback

The original colour version of the images in the plate section are available for
download at www.cambridge.org/9780521008631

Contents

Introduction   *page* 1

1   Volcanism on Earth   4

Why is the Earth hot?   4

Core, mantle and crust   7

Hot spots   11

Plate tectonics   12

Under the volcano   15

The role of gas   18

Eruption!   21

Volcanoes and climate   24

Volcanoes and death   26

Volcanoes and life   30

2   A tour of terrestrial volcanoes   33

Krafla (Iceland)   33

Kilauea (Hawaii)   40

Etna (Sicily)   46

Ol Doinyo Lengai (Tanzania)   54

Mount Pelée (Martinique)   59

3   Volcanism on the Moon   64

First impressions   64

Close-up views of the Moon   66

Apollo 11: touchdown in a lava field   67

Apollo 12: the Ocean of Storms   70

Apollo 15: excursion to Hadley Rille   72

Apollo 17: a geologist on the Moon   76

Lava fountains on the Moon   79

Treasures of *Apollo* 82

Remote-sensing: the big picture 85

The magma ocean 89

Mare volcanism 92

Looking for volcanoes 94

4 A tour of lunar volcanoes 97

Hadley Rille 97

Taurus-Littrow valley 104

The Marius Hills 110

The Gruithuisen domes 116

The Aristarchus plateau and Schroeter's valley 120

5 Volcanism on Mars 127

The discovery of Mars 129

Giant volcanoes 132

Tharsis and Alba Patera 134

Elysium and Hellas 136

Eruptions on Mars 138

Landfall on Mars: the *Viking* missions 143

Martian minerals 145

Pathfinder and the andesite mystery 148

Mars Global Surveyor 150

Spirit and *Opportunity* 151

Meteorites from Mars 154

Young lava 158

Hydrothermal activity and life on Mars 161

Plate tectonics on Mars 165

Volcanoes and the Martian climate 167

Exploring Mars 170

6 A tour of Martian volcanoes 174

Olympus Mons 174

Arsia Mons 184

Alba Patera 190

Hecates Tholus 196

Hadriaca Patera 202

7 Volcanism on Venus 207

An original planet 207

Descent into hell 209

The basalts of Venus 210

Volcanic gases 213

Venus drops her veil 215

In search of plate tectonics 217

Magellan explores Venus 219

Large volcanoes on Venus 221

Spiders and crowns 224

Pancakes and ticks 227

Lava flows on Venus 229

Eruptions on Venus 232

Spatial distribution of volcanoes 234

The life and times of Venus 236

Melting the surface 239

8 A tour of Venusian volcanoes 242

Theia Mons (Beta Regio) 242

Sapas Mons 248

Gula Mons and Idem-Kuva Corona 255

Pancake domes (Alpha Regio) 260

Ammavaru and Mylitta Fluctus 265

9 Volcanism on Io 271

Eruptions on Io 272

The power of tides 275

Persistent hot spots 276

Calderas on Io 279

The color of sulfur 281

The heat rises   285
Galileo at Jupiter: Io revisited   287
The Pillan eruption   289
Global Io   292
The I-24 flyby: Pele's lava lake revealed   295
The I-25 flyby: Tvashtar's curtain of fire   297
The I-27 flyby: calderas and lava flows   298
The 2001 flybys   302
Farewell to Io   304

10 A tour of Ionian volcanoes   305
Pele   305
Loki   312
Prometheus   317
Ra Patera   322
Emakong Patera   326

Bibliography (and websites)   331
Glossary   337
Index   346

Plates between pages 150 and 151*

List of boxed text

Distribution of volcanoes on Earth   8
Mid-ocean ridges   13
The great calderas   19
Hot spots and flood basalts   29

*The original colour version of these plates are available for download at www.cambridge.org/9780521008631

Introduction

Worlds on Fire takes the reader on a tour of the most spectacular volcanoes in the Solar System. Our odyssey starts logically here on Earth in order to lay the foundations of volcanology. We propose field trips to five of our most representative volcanoes, in order to illustrate the range of eruptive behavior, before turning our sights to other planets.

The structure of the book follows this simple pattern: an introductory text presents the discovery and volcanic "flavor" of each planet (odd-numbered chapters), while a field-trip section describes five landmark volcanoes with proposed itineraries (even-numbered chapters). These hiking sections introduce a fair amount of poetic license, as we follow the footsteps of the *Apollo* astronauts to Hadley Rille, climb the mighty slopes of Olympus Mons on Mars, and drift over the searing landscapes of Venus in a balloon. On Jupiter's fiery moon Io, we cautiously approach erupting jets of sulfur and fountains of ultra-hot lava.

Worlds on Fire is written for the layman, but explores volcanic processes in sufficient detail to serve as a guide book for students of the Earth sciences. The text and illustrations include results from the most recent probes: *Mars Global Surveyor*, *Spirit* and *Opportunity* on Mars, and *Galileo* around Jupiter. The language is kept simple and a glossary at the end of the book defines the more specialized terms sometimes used.

I encourage the reader to start with Chapter 1 to review the basics of volcanism and grasp the concepts that will prove handy in later chapters. It is perhaps the most challenging and dense chapter, so bear with me! But it is also wise to skip sections that appear too detailed or complex; each reader should pick the level of reading that fits him or her best.

The field-trip chapters cover 25 different destinations. One way to enjoy this book is to skip around from volcano to volcano, without necessarily reading the introductory chapters. This reading method provides a condensed and hopefully enjoyable picture of a planet and its volcanic activity.

The knowledge presented in this book has been amassed by scores of planetary scientists. My role consisted in summarizing this body of work for the layman and weaving a story line that assembles so many discoveries and themes into a coherent story. A bibliography is listed at the back of the book. This selection of articles is kept short; it is only a sample of the corpus of research material that is available to interested readers and should serve as gateways to some of the journals that specialize in planetary geology. Likewise, the list of websites is the tip of the iceberg for those who want to surf across the Solar System.

I am indebted to the distinguished reviewers who took the time to read and correct my planetary ramblings and in particular my earlier work, *Volcanoes of the Solar System*, which I use as a starting point for this new book. Therefore, I would like to renew my gratitude to Dr. David Clague, former Scientist-in-charge of the Hawaiian Volcano Observatory; Dr. Harrison Schmitt, lunar module pilot and geologist of *Apollo 17*; Dr. James W. Head III of Brown University; Dr. Sean Solomon of the Carnegie Institution; Drs. Larry Crumpler and Jayne Aubele of New Mexico's Museum of Natural History and Science; Dr. Alfred McEwen of the University of Arizona and Dr. Ronald Greeley of Arizona State University.

I am most indebted to Dr. William Hartmann of the Planetary Science Institute in Tucson for going over the Martian chapters of the present book and checking the most recent data that I discuss, as well as to Rosaly Lopes of the Jet Propulsion Laboratory, who did the same with my chapters on Io and the most recent findings from the *Galileo* probe. They provided me with enlightening suggestions and corrections. The errors and subjective slants that remain are mine only.

Most of the illustrations in this book were graciously provided by learning and research institutions, and by individual scientists and their publishers. Last, but not least, I would like to thank Pascal Lee of the SETI Institute for suggesting the title of this book: *Worlds on Fire*.

I hope you will sense the passion and dedication that so many people invested in this book. Grab your spacesuit and your helmet and *bon voyage!*

I   Volcanism on Earth

Volcanoes are fascinating. They are aesthetic landscapes as well as a startling reminder of the forces pent up inside the Earth. Erupting fissures and vents spew out molten rock at temperatures topping 1000 °C, and craters belch plumes of ash that can rise high into the stratosphere and radically alter the climate. Volcanism is also the source of mineral ores, the main provider of gases to the atmosphere, and a leading influence in the creation and evolution of life on Earth.

Because of their wide range in chemistry and eruptive behavior, volcanoes come in a variety of shapes and sizes. Elegant cones crowned by summit craters rise along the Pacific rim from Indonesia and Japan to the Aleutian Islands, Cascades, Central America and the Andes. But volcanoes on Earth also include steep domes, giant shields, and fields of small cones that pepper vast areas of the continents and the deep ocean floor. Volcanoes erupt in the open, or below several kilometers of water, or even under ice caps.

Nor are volcanoes restrained to our planet. One of the important discoveries of the space age is the pervasive nature of volcanism throughout the Solar System. Lava fields, shields and domes show up on the Moon, Mars and Venus. Churning lakes of magma and jets of sulfur are active on Io, Jupiter's fiery moon. These new findings have greatly broadened our perspective on volcanism.

Volcanoes serve a purpose. They release the heat pent up inside planets. Volcanoes act as radiators, circulating molten rock, sulfur or hot water through their plumbing systems – a process that carries the calories up to the surface.

WHY IS THE EARTH HOT?

Planets are born hot. The Solar System came into existence through the gravitational collapse of a cloud of gas and dust, four and a half

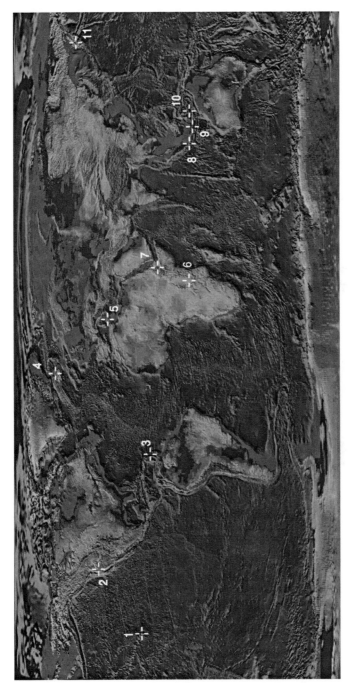

FIGURE 1.1 Topographic map of Earth. Submarine volcanic rifts snake down the Atlantic and the East Pacific. Volcanic island arcs (subduction zones) are prominent in the West Pacific. The major volcanoes discussed or illustrated in the text are indicated by numbers. (1) Kilauea; (2) Mount Saint Helens; (3) Mount Pelée; (4) Krafla; (5) Etna; (6) Ol Doinyo Lengai; (7) Erta Ale; (8) Krakatao; (9) Merapi; (10) Tambora; (11) Kamchatka volcanoes. Credit: JPL-Caltech, U. S. Geological Survey.

billion years ago. The heart of the collapsing system became the Sun. Around the new-born star, the leftover matter collected into swirling lanes of debris – nuggets of rock and ice that slammed into each other to ultimately form the planets. This *accretion* phase was highly energetic, each new impact bringing a blast of heat to the growing planet. Most of this heat was radiated back into space, but a substantial amount remained trapped in the growing body, as layer upon layer of hot debris piled up at the surface. Through this process a planet like the Earth might well have reached a temperature in excess of 2000 K (around 2000 °C) as it reached its adult size in a few tens of millions of years. Its outer layers were probably entirely molten by the heat in a glowing "magma ocean" hundreds of kilometers deep. One giant impact in particular – that of another burgeoning planet – blasted the Earth early in its history and injected a bonus of energy into its deepest layers. As we shall see in Chapter 4, material that was flung into space by this giant collision rapidly coalesced in Earth orbit to form the Moon.

FIGURE I.2 Mount Etna (Sicily), photographed by astronauts aboard the International Space Station. Volcanoes evacuate the heat pent up inside planets. Credit: NASA/JSC.

Planetary accretion is the most spectacular process that heated up the Earth. But simultaneously, a second heating process provided its share of calories – an inner form of gravitational energy. Because the Earth was partially molten, different elements like iron and lead, sodium and potassium, behaved according to their density and buoyancy, some rising to the surface while others sank towards the center. Any sinking body releases energy as it drops from a high level to a lower level within a gravitational field, and the sinking of dense elements like iron and lead released considerable amounts of heat as they trickled towards the center. This should have raised the planet's temperature by an additional 1000 K or so, above the 2000 K already reached by the accretion process. Besides these "outer and inner forms" of gravitational energy, there was a third source of planetary heat: radioactive decay. Elements like uranium are unstable and break up over time into lighter atoms, releasing energy in the process. This nuclear form of heating was particularly efficient in the early days of the Earth, when radioactive atoms were plentiful and decayed readily, such as the highly unstable aluminum-26 that expended half of its stock (known as the element's *half-life*) in the first few million years of Earth history.

Other elements tick away at a slower rate, like potassium-40. Its numbers are cut in half in about 1.3 billion years, so that only 10% of this radioactive "fuel" is still left inside the Earth today. With half-lives of 4.5 and 12 billion years, uranium-238 and thorium-232 have expended only one half and less than one quarter of their stock respectively, so that they still play a major role in the heating of the planet. In fact, 80% of the heat reaching the Earth's surface today is due to radioactive decay and 20% to the accretion of the planet and the sinking of the core, slowly distilled over the aeons.

CORE, MANTLE AND CRUST

Because the minerals that make up the Earth are remarkable insulators – we use rock wool to insulate our homes – internal heat travels very slowly to the Earth's surface. This heat stacks up because of the

Distribution of volcanoes on Earth

There are approximately 1500 volcanoes on Earth that have erupted since the dawn of civilization, 10000 years ago [see the Global Volcanism Network of the Smithsonian Institute]. Of these, 550 volcanoes erupted at least once over the past 2000 years.

Today, an average of 20 volcanoes are erupting around the globe at any one time (above sea level), including 17 that have been doing so semi-permanently over the past 20 years. This select list includes Etna and Stromboli in Sicily, Arenal in Costa Rica, Sakura-Jima in Japan, Semeru in Indonesia, Erta Ale in Ethiopia, Erebus in Antarctica, Kilauea in Hawaii. The list of "old faithfuls" is topped by "surprise" eruptions from other volcanoes (3 or 4 occurrences across the globe at any one time). Over a one-year interval, 60 different volcanoes erupt.

The numbers of volcanoes on Earth grows especially large when one considers the small cones and domes that make up "volcanic fields." A few hundreds of meters tall and several hundred meters in diameter, they are clustered in groups of several hundreds (for example, the Springerville volcano complex, Arizona: 409 vents). There are tens of thousands of these "midgets." On the ocean floor, they are especially abundant in the central rift valley of the mid-ocean ridges, where they stretch out in long chains. Their numbers reach in the millions.

The magnitude of eruptions is inversely proportional to their frequency. Small events are the most frequent. Several eruptions per year emit $0.01\,km^3$ of lava or ash, but only one in a decade reaches $1\,km^3$ (such as Mount Saint Helens in 1981). An eruption of $10\,km^3$ occurs on the average once a century (Krakatao in 1883, Katmai in 1912) and one of $100\,km^3$ once in a millenium (Tambora in 1815). Great ignimbrite eruptions on the scale of $1000\,km^3$ are expected only once every 100000 years (the last one created the Toba caldera, Indonesia, 64000 years ago).

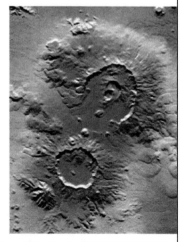

Two volcanoes in the East African Rift. Credit: NASA/JPL/NiMA.

imbalance between heat production inside a planet and heat release at the surface. Heat production is proportional to the quantity of radioactive matter locked up inside a planet and thus to its volume. On the other hand, the escape of heat at the surface is proportional to the planet's area. For a planet with a large radius, like the Earth, the discrepancy between its volume and its area is large, and the cooling cannot keep pace with the heating. The temperature rises. Ultimately, it can reach the planet's melting point.

One would expect the deepest layers of the Earth to be the hottest, and indeed they are. But they are not necessarily molten. This is because temperature is not the only factor involved; pressure also comes into play, as does the composition of the heated material.

The balance of factors is complex and leads to a stratified

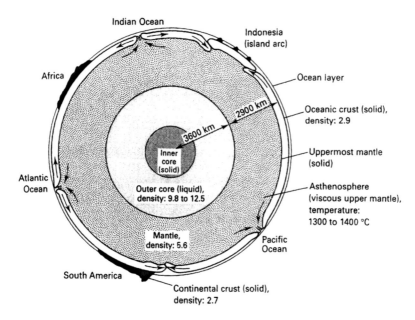

FIGURE 1.3 Internal structure of the Earth. Soon after its accretion, the planet segregated into layers of contrasting compositions and densities: heavy iron sank to form the core, while silicates ended up in the mantle, capped by a thin crust. Credit: adapted from La Documentation Française, Documentation photographique 6107, Les Volcans, 1990.

planet, divided from bottom to top into a core, mantle and crust. Although we do not have access directly to the lower layers – our deepest mines reach only 3 km below the surface and our deepest drill holes a mere 12 km – an indirect assessment can be made through earthquakes. Seismic waves that travel through the planet experience variations in amplitude, speed and direction as they cross through layers of different composition and physical state. These numbers, compiled with other clues, allow us to sketch out a cross-section of the Earth.

At the center of the planet, seismic data reveal a solid *inner core* of iron, perhaps mixed with some nickel or sulfur. The temperature reaches 6000 K (about 6000 °C), but the pressure is so great that melting cannot take place. This inner core extends from the center of the Earth (6378 km) to a specific level, 5150 km from the surface, where the pressure has fallen off faster than temperature and allows melting to occur. Here begins an *outer core* of liquid iron – the only truly liquid layer inside the Earth. Electrical currents flow through this shell of liquid iron and generate the powerful magnetic field that we experience at the surface. This liquid iron shell terminates abruptly 2890 km from the surface, where it is overlaid by a layer richer in silicon, magnesium and aluminum: a *mantle* of interlocked minerals. There is a significant drop in temperature across the boundary – unpoetically labeled the *D-prime* zone – so that the lower mantle is solid. The temperature at the base of the mantle is 2500 K (2200 °C).

Up to the surface, the subtle interplay of pressure and temperature keeps the mantle solid, except in a zone approximately 200 to 100 km below the surface, where the fast-declining pressure lets the mantle adopt a mushy state, with drops of melt collecting along mineral boundaries. This zone is named the *asthenosphere* ("weak sphere" in Greek). It is the prime region where rock melt is generated, which is called *magma*.

Over the aeons, a large amount of magma has risen to the surface, erupted, and cooled to form a solid outer shell, a few tens of kilometers thick. This is the Earth's *crust* on which we live.

HOT SPOTS

So far, we have painted a rather static picture of the Earth. But material within the globe is in motion, because of differences in temperature and density. In particular, the hot core heats the lower mantle, since it is 1000 K hotter. "Blobs" from the lower mantle soak up the heat, expand, and rise through their cooler surroundings like hot balloons through the air. These vertical currents of hot mantle rock are named *plumes* and can rise all the way to the base of the crust. They are typically 100 km in diameter. As they reach shallower levels, mantle plumes have lost little heat during the rise but experience a significant drop in pressure, a combination that causes the rock to melt. The magma released from a plume might ultimately reach the surface, where it produces *hot spot* volcanism. Eruptions from a mantle plume might build large shield volcanoes, as in Hawaii, or spread great sheets of lava across the plains, as happened on the Columbia River Plateau.

FIGURE 1.4 The islands of Hawaii were built over a hot spot in the Central Pacific. The plume of magma from the deep mantle is presently feeding the volcanoes of the Big Island: Mauna Loa and Kilauea. The Pu'u O'o vent (this picture) has been steadily erupting since 1983. Credit: U. S. Geological Survey.

Rising plumes are not the only departures from equilibrium inside the mantle. While hot material rises, cooler and denser rock tends to sink. This is particularly true of ancient crust that has been stacked so thick that its lower layers are pushed down into the mantle. As the growing pressure transforms their minerals into more compact forms, the crust grows denser and begins to sink. As it does, the surrounding mantle flows in to fill the void.

Within the Earth's mantle, this process has become an intricate pattern, with currents of hot rock slowly rising to the surface, releasing their heat, contracting and sinking back down. This looping motion is called *convection*. We are familiar with it in everyday life, for example, when we boil water over a stove. Streams of hot water rise from the bottom of the kettle (usually at the center over the heat source), release heat at the surface and sink back to the bottom along the cooler sides of the kettle. So it is for the hot plastic rock inside the Earth, albeit at a very slow pace: one or two centimeters per year – a couple of meters in a human lifetime. These slow eddies that play under the rigid crust of the Earth are the driving force behind *plate tectonics*.

PLATE TECTONICS

Plate tectonics – now a household name in the Earth sciences – is a global model devised in the 1960s and 70s that encompasses most geological phenomena observed on the planet. It explains earthquakes and eruptions, mountain building, and even the arrangement of ocean basins and continents around the globe.

In the model of plate tectonics, the solid crust of the Earth can tear apart. It grows thinner in places, in response to horizontal stretching, and hot mantle rises to fill the void. Vertical sheets of magma are injected into the overlying crust. Hence, the Earth's crust grows like a deck of cards that is constantly replenished with new cards up the middle of the deck. This process takes place in the middle of ocean basins, basins that grow wider and shove the continents apart on either side – the process of *continental drift*.

Mid-ocean ridges

Mid-ocean ridges harbor the greatest share of volcanic activity on Earth (60% of the magma flux). This activity takes place under 2000 to 3000 m of water, in the form of fissure eruptions with short flows of pillow lavas. Fast-spreading ridges (on the order of 10 cm/year of crustal extension) harbor frequent small eruptions that spread sheets of lava with little relief. Slow-spreading ridges (2 cm/year, as in the Atlantic) harbor less frequent eruptions (once every millenium or so for each location) that build hills several tens of meters high, along the axis of the rift.

Submarines first reached the Mid-Atlantic Ridge in 1974, revealing a wide rift 30 km wide, stepping down 1000 m to an axial valley 3 km wide. Volcanic mounds stretch along the valley and lavas are found to be more than 3000 years old, typical of the intermittent nature of eruptions on "slow" ridges. The faster Pacific ridge erupts much more frequently: in 1993, an undersea robot filmed an eruption on the Juan de Fuca ridge, showing hot water plumes spewing from one end of a 6-km-long fissure, and molten lava flowing out of the opposite end.

Tens of kilometers away from the ridge, there are also chains of large seamounts, 1000 m to 2000 m tall that have the shape of truncated cones with flat tops. Nested craters and calderas up to several kilometers across are comparable to Kilauea crater in Hawaii and likewise reflect the presence of sizeable magma chambers under the volcanic chain. The magma reaches the surface through faults parallel to the ridge axis.

In the Pacific basin alone, there are approximately 10000 large seamounts (taller than 1500 m) and close to one million medium-sized seamounts (taller than 400 m). Small ridge-axis volcanoes (taller than 50 m) are even more numerous: 85 million features are estimated along the Mid-Atlantic Ridge alone. Seamounts and ridge-axis volcanoes constitute by far the largest population of volcanoes on Earth and indeed on any planet known to man.

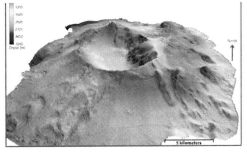

Submarine volcano (West Rota, Pacific Ocean). Credit: NOAA.

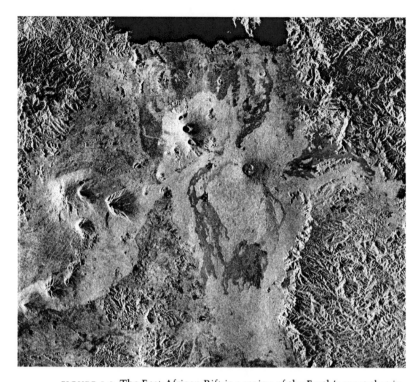

FIGURE 1.5 The East African Rift is a region of the Earth's crust that is being torn by convection currents in the underlying mantle. In the resulting pull-apart basin, magma breaks to the surface, creating chains of volcanoes. Here, the Virunga range is imaged by radar from the Space Shuttle *Endeavour*. Credit: NASA/JPL/JSC.

The newly formed oceanic crust cools as it creeps away from the central heat source (the rising mantle under the ridge). Becoming cooler and denser with time as it rolls out like a conveyor belt, the aging crust ultimately buckles and sinks back into the mantle, setting off eddies around its plunging edge. This sinking process is called *subduction*. The diving of ancient crust fuels a distinct form of volcanism at the surface, which is rich in water (from the water-rich crust) and often explosive. Chains of volcanoes mark the boundaries of sinking plates along the rim of the Pacific basin, notably in the Philippines and Japan, the Cascades of Oregon and the Andes of South America.

These lines of weakness – mid-ocean rift zones where the crust breaks apart and subduction zones where the crust founders and sinks – cut up the surface of the Earth into a mosaic of plates that grow, shrink or shift relative to each other. There are 13 major plates (such as the North American plate, the Eurasian plate and the Pacific plate) and three dozen "microplates" that are caught up like cogs between the larger bodies.

It is believed that 90% of terrestrial volcanism is related to plate tectonics and occurs on the edge of plates: 60% at spreading mid-ocean ridges and 30% over subduction zones. The remaining 10% are the work of hot spot plumes, breaking though the plates in a random fashion. The exploration of the Solar System has taught us that this pattern of volcanism is unique to the Earth. On no other planet does the surface tear apart on a global scale to feed linear chains of volcanoes and jostle plates around like an animated jigsaw puzzle.

UNDER THE VOLCANO

Vertically rising plumes and convection cells bring hot mantle material close to the surface, where the drop in pressure causes it to melt. The composition of the mantle, the proportion of melting of the source rock (5% to 50%), and the circuitous route of the melt up to the surface all introduce a great deal of variability in the lava that is erupted.

When a rock is heated, the more fusible minerals will melt first. Droplets of molten material will trickle out of the source region along mineral boundaries and fine fractures, rising because it is less dense than the encasing solid rock. Droplets collect to form "veinlets" and veinlets merge to form wider seams. In the process the upward flow becomes quicker, because large volumes of magma lose less heat and experience less friction against the conduit walls than smaller blobs of melt.

Magma will ascend as long as it finds pathways to the surface and as long as it experiences buoyancy with respect to its surroundings. Ultimately it will encounter less compact host rock, where the

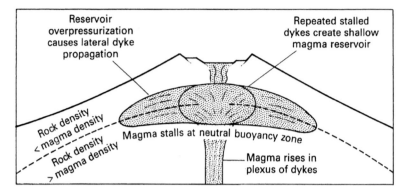

FIGURE 1.6 Development of a shallow magma reservoir in a volcanic edifice. Hot magma rises by buoyancy as long as it is less dense than its surroundings. When this is no longer the case, the magma stalls in a "neutral buoyancy zone," creating a magma chamber and attempting to propagate sideways. Credit: adapted from J. W. Head and L. Wilson, 1992, Magma reservoirs and neutral buoyancy zones . . ., *J. Geophys. Res.*, **97**, 3877–903.

density contrast becomes negligible. The magma will then stall and spread horizontally at this level, which is called the *neutral buoyancy zone*. One such zone is the mantle–crust boundary, several tens of kilometers from the surface in continental settings, and less than ten kilometers under the ocean floor.

These reservoirs are named *magma chambers*. Batches of molten rock slowly cool and solidify in these zones, or find a way to escape. They might then reach the surface or stall again in an upper chamber, just below ground level or even inside a volcanic edifice.

Under Kilauea volcano in Hawaii, for example, a magma chamber is found to lie 2 km to 7 km below the surface. Several branches of molten rock bud out from it, stretching horizontally to form *rift zones*. Kilauea's southeast rift zone extends outward for 15 km, storing a seam of magma. At the distal end, the Pu'u O'o volcano serves as a safety valve and releases lava onto the surface.

The emplacement of magma chambers is often betrayed at the surface by a circular collapse, named a *caldera*. Such structures occur when magma is drained out of the chamber – for instance, when it

FIGURE 1.7 The surface of the Earth collapses over an emptying magma chamber to form a caldera. The pit can later fill with a lava lake, with flows spilling over the rim, as is the case at Erta Ale volcano in the East African Rift. Credit: courtesy of Jean-Louis Cheminée.

migrates along a rift zone to feed an eruption – and the honeycomb of hollow rock is crushed by the weight of the overlying crust. The diameter of the collapse feature is comparable to that of the magma chamber itself. Kilauea's caldera is over 3 km in diameter, with bounding cliff walls 100 meters high. Multiple collapses create nested or overlapping pits, as is the case at Kilauea where the deep Halemaumau pit punches through the southwestern floor of the main caldera.

Besides storing magma, underground reservoirs also transform it chemically and physically, operating as giant stills. As it cools in place, magma will undergo solidification in a progressive, stepwise fashion. Refractory minerals such as olivine will crystallize first and settle at the bottom and on the walls of the chamber to form a natural slag. The remaining liquid will become poorer in iron and magnesium (absconded by the olivine crystals) and relatively richer in silicon, sodium and potassium. As the temperature continues to drop, other crystals form and the remaining liquid keeps evolving towards more alkaline and siliceous chemistry. This process of *fractional crystal-lization* can yield a whole suite of different lavas at the surface if the evolving magma is tapped at different points in time, or at different levels within the stratified chamber.

One common family of lavas is known as the *alkaline suite*. It is found in hot spot settings. The chain of volcanoes in central France (chaîne des Puys) provides a clear example of the process. The first eruptions out of the magma chamber yielded primary lavas named *basalt*. As the underground magma batch evolved over time, the erupted lavas became lighter-colored: poorer in iron and magnesium, and richer in sodium and potassium. They take on the names *leuco-basalt, trachyandesite, trachyte* and *phonolite* as their proportion of silica increases.

Another classic differentiation trend, known as the *calc-alkaline suite*, is typical of subduction zones, where the magma is water-rich. With increasing silica content, the lavas range from *andesitic basalt* to *andesite, dacite* and *rhyolite*.

THE ROLE OF GAS

The outpouring of lava at the surface is often spectacular, because there is a small quantity of gas dissolved in the magma. As it nears the surface, the pressure drop causes the gas to expand dramatically and propel the molten rock high into the air. Only when the gas content is low does the magma remain undisturbed and pour out peacefully from the vent.

Water and carbon dioxide are the two main gases that drive eruptions. In the mantle and lower crust, their molecules are locked up inside minerals like amphibole or carbonate. When partial melting takes place, the volatile elements leave their host minerals and readily enter the liquid phase.

Volatile molecules like water actually help the rock to melt. For example, the melting point of *peridotite* – a rock typical of the upper mantle – is lowered by 300 K when water is present. In fact, if the upper mantle were bone dry, it would not melt at all under the

The great calderas

Some of the largest volcanic eruptions on Earth are thick sheets of welded ash, named ignimbrites. The latest example is the Katmai ignimbrite field in Alaska (1912), also known as "the Valley of 10000 smokes," formed by the eruption of $12\,km^3$ of ash. The eruption of Taupo in New Zealand, in AD 200, created $35\,km^3$ of welded ash. Much larger events occur every few tens of thousands of years: the eruption of Toba in Indonesia, 75000 years ago, released $1500\,km^3$ of ash and left an oval caldera 80 km long and 30 km wide. The largest event identified so far occurred 28 million years ago at La Garita (USA). The eruption expelled $5000\,km^3$ of ash and likewise created a caldera 75 km by 35 km.

An ignimbrite field is caused by the catastrophic emptying of a magma chamber, accompanied by caldera collapse at the surface. The ash is erupted out of ring faults on the periphery of the collapsing structure. Often the collapse occurs piston-style, with the entire floor of the caldera collapsing in one piece. In other cases, the ring faults are incomplete and act as a hinge: the asymmetrical collapse is called a "trapdoor" subsidence.

After a major eruption, magma progressively replenishes the caldera, creating a bulge or "resurgent" dome in the center. Small lava domes also form atop the ring faults. The replenishing of the collapsed chamber with siliceous volatile-rich magma takes on the order of several hundreds of thousands of years, leading up to the next catastrophic eruption. Long Valley caldera in California last erupted 760000 years ago and presently experiences earthquakes around the magma chamber at depth that are closely being monitored.

Primary locations for resurgent calderas and ignimbrite fields are regions of crustal extension, in continental rift zones and the back setting of island arcs: the Basin and Range of North America, Central America, the Andes, and the Western Pacific margin. The number of resurgent calderas identified so far on Earth reaches in the hundreds.

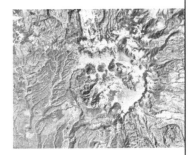

Valles caldera, New Mexico. Credit: NASA.

FIGURE 1.8 Volcanoes in subduction zones are fed by gas-rich magma. Here, the Merapi dome in Indonesia smolders away. Collapse of the dome allows the gas to expand into fiery avalanches. Credit: courtesy of Jacques-Marie Bardintzeff.

temperatures that prevail today. There would be no mid-ocean ridges, no subduction volcanoes and no plate tectonics. Only hot spots would operate, driving abnormally hot rock from the lower mantle to the surface. This might well be the case on Venus, which lacks water and also lacks plate tectonics.

On Earth, "water-fluxing" keeps the mantle melting. Depending on its composition, the rising magma might contain up to 5% water and 1% carbon dioxide by weight, as well as minor amounts of sulfur, chlorine, fluorine and other volatile elements. As the magma rises and the pressure drops, these volatiles come out of solution as small bubbles and separate from the melt. Much carbon dioxide escapes in this way before the magma reaches the surface, but large quantities of water vapor and sulfurous gases remain dissolved in the magma until it is only hundreds of meters from the surface. At this point, the residual gas will cause the melt to foam, and the froth will affect the course of the eruption.

ERUPTION!

When there is only a small amount of gas left in the magma – less than 0.1% by weight – the bubbles are too small to disrupt the melt significantly. The magma erupts as a peaceful lava flow. One effect the trapped gas does have – especially water vapor – is to decrease the viscosity of the mix and help it to flow over longer distances.

If the volatile content in a basaltic melt reaches several tenths of one percent, the number and size of bubbles will increase. Their buoyancy will accelerate the magma up the chimney, from standard velocities of a fraction of a meter per second at depth to blow-out speeds at the vent of 100 m/s and higher, comparable to the exhaust speed of a jet engine.

Once the expanding gas bubbles take up more than three quarters of the volume in the conduit, they blow the magma to shreds. The erupting mixture then takes the form of a roaring lava fountain. Clots of magma fall back around the vent and often coalesce to form

FIGURE 1.9 The eruption of Mount Saint Helens in the Cascades range (USA). After a lateral blast caused by the collapse of the volcano's northern flank, the eruption switched to plinian activity with a column of ash and gas that rose 20000 m. Credit: U. S. Geological Survey, Cascades Volcano Observatory.

lava flows. Typically, fountains spray magma a few hundred meters in the air (up to 800 m for 0.6% water content), but exceptional heights of 1600 m have been reported for the 1986 Izu-Oshima eruption in Japan. This style of activity is known as *hawaiian eruptions*, since it is typical of Hawaiian volcanoes like Kilauea.

There are cases when gas bubbles merge in the magma chimney, especially if the conduit is narrow and constricted. The bubbles rise faster than the magma itself and coalesce into a number of large "slugs." Fountaining is intermittent, occuring only when a slug of gas reaches the surface of the magma column and "pops." This intermittent spray is known as a *strombolian eruption* because it is typical of Stromboli, a volcano of the Eolian islands north of Sicily.

Gas can also become trapped under a plug or cap rock that obstructs the crater, especially if the lava is viscous. As the gas accumulates, the pressure rises until the obstruction bursts and the pressurized gas jets out with great force. The blast sends shock waves through the atmosphere and propels large blocks of cap rock to great distances. Exhaust speeds can become supersonic (over 400 m/s). At Sakurajima volcano in Japan, blocks the size of cars and weighing 20 tons have been tossed 3 km from the crater. These explosive events are named *vulcanian eruptions*, since they were first described at Vulcano crater, also in the Eolian islands.

In some magmas the volatile content is extremely high, even without the damming effect of an obstruction. This is especially the case in subduction zones, where the melting of oceanic crust and sediment flushes large amounts of water vapor and carbon dioxide into the magma. Their proportions can reach 4% and 1% by weight respectively. Rather than creating a short blast, as in a vulcanian eruption, the gas will roar over periods of minutes to hours and expel the entire magma column skyward in a plume of ash that can reach heights of 20 to 30 kilometers. These are named *plinian eruptions*, in reference to the eruption of Mount Vesuvius that buried Pompei in AD 79 and was described in writing by Pliny the younger. In a plinian eruption, the ash rains down over large areas downwind from the volcano.

FIGURE 1.10 Vulcanian eruptions, caused by overpressure in a clogged conduit, can eject large blocks of lava or "bombs." This boulder, weighing several tons, lies on the rim of Vulcano crater in Sicily. Credit: photo by the author.

In cases where the discharge rate is too high, the ash piles up in the atmosphere above the vent and can no longer be lofted by the hot gases: the plume collapses onto the flanks of the volcano. These collapsing clouds create *base surges* that spread layers of ash on the slopes (*pyroclastic deposits*). Similarly, domes of viscous lava can crumble and free their trapped gases in blasts of hot ash that likewise roll down the slopes. Such hot avalanches are named *nuées ardentes* ("fiery clouds") and constitute *pelean eruptions* in memory of the eruption of Mount Pelée in Martinique, that claimed close to 30 000 lives in 1902.

Magma can also mingle with "outside" volatiles, such as groundwater, seawater or even ice. One instance is when rising magma intersects the water table. The water flashes to steam and blasts the overlying rock in a way that closely resembles a nuclear explosion. It creates a shallow crater known as a *maar*, surrounded by a *tuff ring* of fall-out debris.

Lava can also mix with ice or water at the surface and create flash floods of mud that sweep down into the valleys. These mudflows are known as *lahars* (Indonesian term) or *jökullaups* (the Icelandic variety).

VOLCANOES AND CLIMATE

The Earth's atmosphere was initially shaped by volcanoes, which released water vapor, carbon dioxide and sulfur dioxide into the environment. The eruption of large quantities of ash and gas can still affect the atmosphere today and alter the climate. An addition of carbon dioxide, for example, contributes to greenhouse warming. Sulfurous gases induce cooling, because they turn into droplets of sulfuric acid (*aerosols*) that absorb and reflect sunlight, and cut down the amount of heat that reaches the ground. Ash particles also cut off sunlight, but ash is quickly washed out of the atmosphere and its influence is short lived.

In summary, the effect of an eruption on climate is a mixed bag, but most documented cases show a net cooling effect. The first scientist to recognize such a pattern was Benjamin Franklin, who linked the abnormal weather over Europe in 1783–84 to the great Laki eruption in Iceland. During this colossal outpouring of basalt, lava fountains rose a full kilometer into the sky and volcanic gases reached up into the stratosphere. Over 100 million tons of sulfur dioxide were released over a couple of months – as much as our industries release today in a year.

The effect of an eruption on the global climate is most pronounced when the gas plume penetrates the stratosphere, where aerosols can stay suspended up to a year. When the volcano is located close to the Equator, the billowing plume can also spread into both hemispheres. This was notably the case of the two largest eruptions of the nineteenth century: Tambora (1815) and Krakatao (1883), both in Indonesia.

In the case of Tambora, the output of sulfur dioxide was comparable to the one at Laki: over 100 million tons (MT). The tempera-

FIGURE 1.11 An ash-laden volcanic plume extends out to sea from an erupting volcano on the Kamchatka peninsula, as viewed from the Space Shuttle (STS-68 mission, 1994). Volcanoes contribute large amounts of volatiles to the Earth's atmosphere – principally water vapor and carbon dioxide. Credit: NASA, courtesy of Debra Dodds, Johnson Space Center.

ture dropped an estimated one to two kelvins in New England, where 1816 was known as "the year without a summer" (it snowed in July). World wide, the temperature drop was close to 0.5 K.

The 1883 Krakatao eruption was half as potent (50 MT of SO_2) and world temperatures decreased by 0.3 K. A similar drop was observed after the 1991 eruption of Mount Pinatubo in the Philippines, when 20 MT of SO_2 were propelled into the stratosphere.

A temperature change of a fraction of a kelvin does not sound like a lot, but over an entire year it represents a substantial amount of heat subtracted from the Earth's budget. Locally it can lead to serious perturbations and have repercussions on food crops and human health. The harsh winter that followed the Laki eruption in 1783 played a role in the poor harvests and the famines that led up to the French revolution. Chinese records mention similar droughts and freezing events in the wake of major eruptions, notably those of Thera-Santorini (1620 BC) and Mount Etna (44 BC), both in the Mediterranean. Large historical eruptions have thus spewed hundreds of millions of tons of gas into the atmosphere and led to cooling anomalies on the order of a couple of kelvins. But on longer time scales, gigantic eruptions occurred that dwarf anything witnessed by humans. The Toba eruption in the Philippines, 74 000 years ago, emitted 20 times more ash and gas than the largest event in history (which was Tambora in 1815). Its effect on climate is believed to have been much worse. Perhaps the very history of *Homo sapiens* was affected by the event. According to DNA extrapolations, it appears that our species underwent a drastic reduction in numbers around that period. Scaling up the magnitude of eruptions, can a threshold be reached where the entire biosphere is at risk?

VOLCANOES AND DEATH

Volcanic eruptions can be harmful to life on a local scale. Lava flows are seldom fatal, except when they reach exceptional velocities. There is one rare case, in 1977, when a super-fluid flow from Nyiragongo volcano in Kenya rushed down the slopes at 15 km/h and overcame a village (70 casualties). Bursts of gas can suffocate entire communities, as happened in 1984 at Lake Nyos in Cameroun (2000 casualties).

The greatest hazard to life comes in the form of ash falls, base surges, nuées ardentes and mudflows that reach populated areas at much greater speeds. The hot cloud of ash that rolled down the slopes of Mount Pelée in Martinique and claimed 28 000 lives on

FIGURE 1.12 The catastrophic eruption of Tambora volcano in Indonesia, in 1815, decapitated the summit of the volcano, leaving a circular caldera in the middle of the island. This most energetic eruption of the past millenium caused over 100000 casualties, due to the tidal waves, famine and disease that resulted. Credit: NASA.

May 8, 1902, is the most tragic example, but there have been many others.

Gas and ash can also cause the poisoning of pastureland and food crops, as they did during the Laki eruption in 1783. The high concentration of chlorine and fluorine and the gritty ash covering the grass killed most of the sheep and starved to death a quarter of the Icelandic population. Famines are often compounded by the spreading of diseases such as typhus and cholera, as happened in the aftermath of the Tambora eruption in 1815, when the death toll reached over 100000 people.

Paleontologists who study the fossil record have also speculated that extraordinary eruptions – the likes of which only occur every few million years – alter the climate on such a grand scale that they lead to the elimination of entire plant and animal species – massacres known as *mass extinctions*. They point out that several of the greatest mass extinctions on record took place during periods of extensive volcanic activity. This was the case of the End Cretaceous crisis, 65 million years ago, that coincides with the eruption of the Deccan Traps in India – one of the half-dozen great igneous provinces on Earth. The End Permian crisis, 250 million years ago, likewise coincides with the eruption of the Siberian Traps.

The proposed scenario is that these eruptions injected massive amounts of ash and sulfurous aerosols into the stratosphere, causing a "volcanic winter" – a scaled-up version of the climatic perturbations that Laki demonstrated in 1783. Large eruptions that build up the trap lava fields are believed to surpass the discharge rate measured at Laki, and do so for years rather than months (the Roza flow field of the Columbia River Plateau totals $700\,km^3$, versus $15\,km^3$ for the Laki flow field). However, it remains to be proven that such eruptions, separated by years of quiet, disturb the climate to the point of eliminating scores of species worldwide.

Critics point out that increasing the amount of ash and aerosols in the stratosphere does not increase the atmospheric disturbance in the same proportions. A level is reached when particles begin to clump together into large flakes that are less obscuring as a whole than the finer material. Bigger and heavier clumps also fall out of the atmosphere much faster. So in essence, more is less.

The lethal consequences of large eruptions are thus unclear. So is their relationship with mass extinctions in the first place. When examined in detail, there is indeed no compelling evidence that any of the mass extinctions were caused by volcanic eruptions. In fact, the flagship case – the End Cretaceous mass extinction that saw the end of the dinosaurs – is not supported by any of the chemical and mineralogical clues found in the clay layer (the "K–T" boundary) that marks the mass extinction. The presence of rare metals like iridium and of mineral grains shocked at high pressure are impossible to explain by volcanism, despite a few claims made in the 1980s. These clues are symptomatic instead of a large asteroid impact that created an infernal heat blast across the entire planet (caused by the ejecta fall-back) and an opaque cloud of soot and dust that dwarfed any volcanic plume imaginable.

Hot spots and flood basalts

Hot spots are places on Earth where plumes of hotter-than-average mantle rock rise to the surface. Such plumes are usually shaped like inverted drops – typically 100 km in diameter – with a large head and a tapering tail. Most are thought to originate at the core–mantle boundary, 2700 km under the surface, and are about 200 K (200 °C) hotter than their surroundings.

When they "break" at the surface, plume heads create abundant volcanism for hundreds of thousands of years. The resulting lava fields are known as Large Igneous Provinces, flood basalts, or *traps* (from the Scandinavian word for "stairs," because the lava layers appear as great steps in the landscape).

There have been about 10 episodes of these monumental eruptions over the past 250 million years, starting with the Siberian Traps that encompass close to 2 million cubic kilometers of lava and pyroclastic material, erupted in perhaps less than a million years. Other famous examples include the Parana-Etendeka Traps of South America (132 million years ago) and the Deccan Traps of India (65 million years ago) that are equally voluminous. The latest occurrence is the relatively smaller and well-studied Columbia River Plateau (17 million years ago) in the American Northwest.

Traps are commonly made of basalt, but basaltic andesite and andesite were erupted by the Columbia River Traps. In this field, individual flows are up to 600 km long (Pomona flow) and 1300 km³ in volume (Roza flow) and testify to a high eruptive rate or at least an efficient isolation of the hot lava in tunnels and tubes. Trap eruptions probably release large volumes of gas (notably SO_2) into the atmosphere and are held responsible by some reviewers of global climate upheavals and mass extinctions of the biosphere (see text for a critical review of this theory).

Lava plateau in Argentina (radar image). Credit: NASA/JPL/JSC.

The impact and its lethal effects are confirmed by the discovery of the collision scar (Chicxulub crater in Mexico) and of soot concentrated in the clay layer that amounts to the combustion of at least half the biosphere. The extinction of species at the K–T boundary is proven to have been nearly instantaneous, whereas the volcanic episode of the Deccan Traps were drawn out over hundreds of thousands or perhaps millions of years. Proponents of the volcanic theory still claim that the Deccan eruptions weakened the biosphere prior to the impact – and thus contributed to the mass extinction – although again there is no proof to that effect.

Likewise, other large volcanic episodes that are synchronous with mass extinctions – such as the Siberian Traps and the End Permian crisis – have left no clues that support a causal relationship. There are also cases when volcanic episodes of similar magnitude are not associated with any mass extinctions, such as the Karoo and the Parana-Edenteka Traps, which erupted respectively 183 and 132 million years ago.

VOLCANOES AND LIFE

Large volcanic events are not convincingly linked with global-scale destruction of the biosphere. On the other hand, there is some indication that volcanism is linked to the creation of life on Earth.

Deep-sea diving on the mid-ocean ridges has uncovered an extraordinary menagerie of original life forms in the hot waters springing out of the seafloor. Crab and shrimp, clams and tubeworms colonize the warm lava flows and the hydrothermal vents. But it is the microbial community that is the most astonishing. Vents on the seafloor emit hot water plumes laden with white flocculent matter: flakes of sulfur produced underground by sulfide-oxidizing bacteria.

Bacteria that thrive in extreme environments – high temperature and high pressure in this case – are named *extremophiles*. Those of the mid-ocean ridges do not exploit sunlight or decaying organisms to fuel their metabolism, but tap instead the metallic ions carried in the hydrothermal waters. Some bacteria reduce sulfate ions, while

FIGURE 1.13 Active hydrothermal vent at a depth of 3000 m on the East Pacific Rise (1982 Cyatherm exploration campaign). This hot spring spews forth 300 °C fluids rich in dissolved minerals that create a haven for extremophile bacteria and a complex food chain. Credit: © Ifremer.

others reduce iron or manganese oxides. The mid-ocean ridge environment provides both the chemicals – the metals leached from the lava – and the heat gradient necessary to exploit them (mixing of hot and cold water).

Mid-ocean ridge bacteria belong to the newly-defined class of *archebacteria* – a wholly separate kingdom of life, distinct from other bacteria and nucleated cells. Upon their discovery, it was first believed that archebacteria were the most primitive life forms that preceded all other bacteria on the tree of evolution. It appears today that they are cousins rather than ancestors of the more common types. The fact remains that some of the earliest forms of life thrived in hot volcanic environments, which raises the possibility that life itself originated in such a context. Considering the constant battering of the early Earth by asteroids and comets, the deep seafloor would have been a haven for the development of living organisms. Instead of presiding over the death of species, volcanism might well have ushered in the adventure of life on Earth.

This new line of research has far-reaching implications for the search for life on other planets as well. Early in its history, Mars experienced abundant volcanism and hydrothermal circulation. When we set out to look for life – extant or fossil – on the red planet, volcanoes will rank among the most promising sites to explore.

But before we embark on our tour of the Solar System, let us visit a select number of volcanoes on Earth which will illustrate the various forms of activity that occur on our home planet and that we might encounter on other worlds.

2 A tour of terrestrial volcanoes

Krafla (Iceland)

Latitude:	65° 45′ N
Longitude:	16° 45′ W
Length of rift:	90 km
Width of rift:	up to 8 km
Altitude:	500 m
Relief:	3000 m from the seafloor
Age:	<2 million years
Caldera:	10 km × 8 km
Depth:	filled with lava

FIGURE 2.1 Krafla 1980 vent and pahoehoe flow. Credit: photo by the author.

Most volcanic activity on Earth takes place on mid-ocean ridges, 3000 meters below sea level. Fortunately, unusual activity in the North Atlantic – the contribution of a mantle plume – has led to the build up and emergence of the Mid-Atlantic Ridge to form Iceland, a large volcanic island between Greenland and the British Isles.

In Iceland, mid-ocean ridge volcanism can be studied above sea level, although the subaerial setting changes some of its characteristics. Rather than stacking up around the vents as water-chilled "pillows," the basaltic lava spills out over great distances. The "dry ridge" is best exposed in the northern region of Lake Myvatn, where it strikes inland, north–south from the coast. The 100-km-wide swath of recent lava is subdivided into four sections, each characterized by a central volcano complex and associated fissure swarms.

Krafla is one such complex and exemplifies rift-style volcanism. Its topographic expression is that of a rift 3 km to 8 km wide, with

FIGURE 2.2 Iceland viewed by the *Terra* satellite (Krafla shown by arrow). Credit: Jacques Descloitres, MODIS Rapid Response Team, NASA/GSFC.

FIGURE 2.3 The Viti explosion crater (maar). Credit: photo by the author.

bounding walls less than 30 m high. It stretches over a length of approximately 90 km and is characterized by fissure swarms and occasional ridges of lava and ash. At the center of the complex, a subdued caldera 8 km in diameter, filled in by its own lava flows, overlies a shallow magma chamber.

In historical times, the Krafla complex has erupted once every 300 to 400 years on average: around AD 850, AD 1300, in 1724–29, and in 1975–84. During the most recent cycle (1975–84), the stretching of the crust reached eight meters. Combining this figure with a 400-year rest period yields an average spreading rate of 2 cm/year, typical of the Mid-Atlantic Ridge. When fractures break open, magma wells up from the underlying mantle reservoir. Most of it congeals in the fissure zone without reaching the surface. During the 1975–84 eruption, one cubic kilometer of fresh magma was emplaced in the rift zone, but only a quarter (0.25 km³) was erupted at the surface. Eruptions took place along fissure segments 2 km to 8 km in length, with fountains of lava spouting skyward and ribbons of lava spreading onto the surface. Such outbreaks took place once in 1975, twice in 1977, three times in 1980, twice in 1981 and once in 1984, each phase lasting only several days.

The mineral make-up of lavas at Krafla indicate that the magma source is the upper mantle, at a depth of about 10 km. Outside the central caldera, the magma ascends directly to the surface without changing its composition: the erupted lava is tholeiitic basalt, characteristic of mid-ocean ridges. Below the central caldera, on the other hand, the magma stalls in a crustal reservoir, 2 km below the surface, where it decants, precipitates crystals and changes composition in the process. With progressive silica enrichment, it yields andesite, dacite and rhyolite. This siliceous "cream" is erupted at the surface each time a new batch of magma ascends from the mantle into the chamber and provides the extra heat and gas to expel it to the surface.

The presence of a shallow magma chamber also promotes hydrothermal circulation. Snowmelt and rainwater that infiltrate the system are turned into hot springs. Locally, they alter the ground to yield colorful clays and boiling mud pots. The geothermal field is exploited at a power plant on the southern edge of the caldera. Kilometer-deep bore holes tap hot water and steam that drive turbines. The generators have a maximum output of 60 megawatts of power.

Krafla is therefore a showcase of mid-ocean volcanism, magmatic differentiation, hydrothermal systems (including extremophile biology), and geothermal energy. It also yields a variety of landscapes that serve as "analogs" to better understand features on other planets, especially Mars. The following field trip highlights some of these analogue landscapes and takes the hiker into the heart of the Krafla volcanic system.

Itinerary

The best time to visit Iceland, which lies just below the Arctic Circle, is June to October. Plan for mild but rainy and windy weather. From the capital Reykjavik, the Krafla volcanic rift is reached by a 40-minute domestic flight to the northern town of Akureyri and a 2-hour

FIGURE 2.4 Vent of the 1980 eruption with lava flows in the
foreground. Credit: photo by the author.

car drive east to Lake Myvatn and to the village of Reykjahlid. Hotels
and bed-and-breakfasts can be found in both towns.

A visit of the area should start with a look around the lake's
shore, where Krafla's 1724–29 lava flows crept over the marshy
terrain, setting off steam explosions that created tuff cones up to 100
meters in diameter. These are textbook examples of *pseudocraters*,
also called "rootless vents." Similar features detected on close-up
imagery of Mars hint to a similar type of interaction between lava and
water-rich ground.

Martian and lunar landscapes are plentiful in the Myvatn area.
Overlooking the lake, the Hverfjall tuff cone, 1200 m in diameter and
160 m tall, is well worth climbing. Nested inside its shallow crater, a
small cone gives it the appearance of an impact crater with its central
peak. No wonder the *Apollo* astronauts trained in the area before their
moon flights! A couple of kilometers to the southeast rises the steeper
dome of Ludent – an example of differentiated lava (dacite).

Other examples of differentiated lava are found inside the

perimeter of the Krafla caldera. From Reykjahlid, drive five kilometers east (you should stop to look at the boiling mud pots of Namaskard along the way). Branching off to the north, a side road leads to the Krafla geothermal plant, a 5-km drive along the outer flank of the rift zone. After passing the geothermal site – where the extracted steam drives two 30-MW turbines – you reach the edge of the most recent lava field (the flows of 1984 stopped less than one kilometer from the plant). The road ends at a parking lot where visitors can walk up to the edge of Viti crater, which is a steep explosion pit (called a *maar*) that heralded the start of the 1725–29 eruption. The magma that hit the water table at Viti, and caused the *phreatomagmatic* eruption, was a form of rhyolite, a siliceous "cream" squeezed out of the Krafla magma chamber. A lake inside the crater shines blue, due to the high silica content of the waters.

At Krafla, siliceous lavas are principally found around the periphery of the caldera, under which the magma chamber is located. Often, batches of magma from the upper mantle rise directly to the surface, without stalling and decanting in chambers along the way.

FIGURE 2.5 Dikes jutting out of a ridge of hyaloclastites (subglacial eruption). Credit: photo by the author.

This was the case during the 1975–84 eruption, when pristine basalt was erupted. The black lava flows are visible from Viti to the northwest: they cover 66 km² of the rift zone. One can reach their southernmost reaches by following a trail over half a kilometer to a rhyolitic complex with a few mud pots. Dark lava flows encroach on the older terrain and can be traced back to their many vents. Flows of smooth *pahoehoe* lava are common, as are lava tunnels and push-up ridges.

Krafla is thus a remarkable showcase of planetary volcanism, displaying the mechanism of mid-ocean ridges, as well as lava-flow textures, magma differentiation and lava–water interaction. Older volcanic edifices around Krafla also display the interaction of glaciers and volcanoes. Steep-sided edifices were molded inside casings of ice; these forms are called *tuyas* and *hyaloclastic ridges*. Some edifices on Mars have similar shapes. It is therefore no surprise that planetary geologists travel to the Krafla area to study and model Martian volcanism.

Kilauea (Hawaii)

Latitude:	19° 25′ N
Longitude:	155° 17′ W
Area:	1500 km²
Volume:	700 km³ (above water)
Altitude:	1500 m
Relief:	5500 m above seafloor
Age:	<100 000 yrs
	currently active
Caldera:	5 km × 3 km
Depth:	0 to 150 m

FIGURE 2.6 Kilauea caldera: the Halemaumau pit. Credit: U. S. Geological Survey.

Kilauea is a shield volcano that makes up the southeastern section of the Big Island of Hawaii. It grows buttressed against the southern flank of the central Mauna Loa shield and is presently the most active volcano of the island.

The Hawaiian Islands are a chain of volcanoes in the Central

Pacific that owe their existence to a mantle plume piercing the oceanic crust. As the Pacific plate drifts to the northwest above the stationary hot spot, volcanoes are formed overhead and carried away on the plate's back, like hot cakes on a conveyor belt. As they leave the zone of influence of the central plume, the volcanoes are disconnected from their magma source. They shut down and, along with the underpinning crust, cool and contract, ultimately disappearing below sea level. The chain of the Hawaiian Islands thus comprises extinct volcanoes to the northwest (Kauai is 5 million years old), whereas the youngest island in the southeast – the Big Island – is less than one million years old. Within the latter, current activity is likewise focused under its southeastern volcano: Kilauea. Mauna Loa volcano is also still active.

Whereas Mauna Loa has had time to grow to a record volume of $80\,000\,km^3$ above the seafloor – making it the largest volcano on Earth – Kilauea is growing on its southern flank and has not yet reached its maximum size. It is a shield volcano, albeit asymmetrical because Mauna Loa inhibits its growth to the north. Kilauea boasts a central caldera averaging 4 km in width, 1500 m above sea level, and two fissure swarms (10 to 20 km wide) extending from it, roughly parallel to the coast: the southwestern and eastern rift zones.

At Kilauea, the magma originates in the upper mantle, at roughly 50 km depth, and collects in a magma chamber directly under the caldera, 2 to 7 km below the surface. From there, the magma travels underground through the rift zones. Half of the magma congeals underground as intrusions, while the other half erupts at the surface. The extruded lava is *olivine tholeiite* – a basalt typical of the proportionately large amount of melting (25% to 30%) that occurs in a mantle plume. Decanting of this mantle stock in a magma chamber can yield more evolved types of lava: alkaline basalt, hawaiite, mugearite and trachyte (known as the alkaline suite).

On the flanks of Kilauea, 90% of the lava is less than 1100 years old. The basaltic flows are emplaced during frequent short eruptions (often less than a week long), as well as during longer events. A major eruption began in 1983 in the eastern rift zone (Pu'u O'o vent) and has

FIGURE 2.7 Shuttle radar view of Kilauea caldera (center) and the eastern rift zone (right) with lava flows descending to the sea. Credit: NASA/JPL/JSC.

proceeded virtually uninterrupted for more than 20 years. The discharge rate averages 400000 m³ per day – approximately 0.15 km³ per year – which makes Kilauea the most productive volcano on Earth today. Much of the lava travels underground through tubes that insulate it from the air and guarantee long run-out distances.

Kilauea's summit caldera is 5 km × 3 km wide and contains a nested pit named Halemaumau, 1000 m in diameter and presently less than 100 m deep (its depth reached 400 m in the past). The pit episodically hosts a lava lake, as it last did in 1967. The formation of the larger caldera probably dates back to 1790, when the magma chamber collapsed. Powerful steam explosions were associated with the event, spreading a layer of pyroclasts around the summit, known as the Keanakakoi ash. Large eruptions and caldera collapses are believed to occur every 1500 years on average.

Other tectonic motions affect the volcano, as the southern flank of Kilauea stretches and slips seaward under its own weight. A crustal extension of 2 to 3 m took place in the eastern rift zone in January 1983 – fissures that were used by the magma to reach the surface and initiate the current eruption. Larger displacements have occurred in the prehistoric past, with entire sections of the volcano

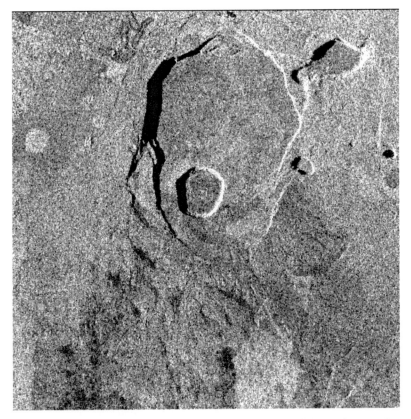

FIGURE 2.8 Shuttle radar view of Kilauea caldera, showing the nested Halemaumau pit in the south and the Kilauea Iki pit to the northeast. Credit: NASA/JPL/JSC.

slipping down slope along major faults. Steep scarps – the *pali* – mark the locations of these mass movements. The Hawaiian Islands are surrounded by giant debris flows on the ocean floor, which attest to major collapse events. They provide an interesting analog to interpret the scarps and hummocky landscape – known as the aureole deposit – which surround Olympus Mons on Mars.

Itinerary

A visit of Kilauea volcano – a one-hour drive from Hilo on the Big Island – begins at the summit crater (a caldera). On the western rim, next to the Hawaiian Volcano Observatory, a museum presents the history and scientific monitoring of the volcano, and provides maps and field guides to the many trails of the Hawaiian Volcanoes National Park.

One can easily spend a day or two exploring the caldera itself and its many wonders. A drive from the Observatory, counter-clockwise around the caldera, first passes across faults and gullies that breach the southwestern rim. They expose outcrops of Keanakakoi ash – the air fall that accompanied the collapse of the summit and the formation of the caldera in AD 1790. The caldera rim is lowest here and constitutes an outlet through which lava can flow out onto the flanks.

The road descends onto the caldera floor and reaches a parking lot on the edge of the inner pit of Halemaumau. The pit occasionally harbors a lava lake, and can also be rocked by phreatomagmatic explosions. Shattered rock from a blast in 1924 litters the periphery of the pit. Sulfur-rich gases vent through the fractures and yellow sulfur deposits can be spotted in the pit wall. Besides the Halemaumau pit, Kilauea caldera hosts numerous lava flows, fissures and spatter cones, which make for instructive hikes. Across the 3-km-wide floor, the caldera walls display lava flows in cross-section, which built up the shield summit, as well as intrusions that punched through the layering (notably the Uwekakuna *laccolith*).

Other pit craters are found on the periphery of the caldera. As we leave the Halemaumau parking lot, the drive climbs back onto the rim, indented by the 300-m-wide Keanakakoi crater (parking pull-out). Its most recent eruption in 1974 spilled lava across the road into Kilauea caldera. A couple of hundred meters down the road, one can hike down the slope to a gully that the fluid lava followed during the eruption. The gully can be viewed as a scaled-down version of a sinuous rille on the Moon. It is lined with a plastering of glassy lava. The benchmark is highest in the bends, where the flow surged upward, carried away by its momentum.

Another "satellite crater" worth a hike is Kilauea-Iki, which last filled with lava in 1959. The 120-m-thick lava pond took 20 years

FIGURE 2.9 Eruption on the eastern rift zone, with a cooling lava pond.
Credit: USGS photo by J. D. Griggs.

to completely solidify, although drill holes show that the temperature
at the bottom of the pile is still close to 1000 °C. The lava is *picritic
tholeiite* basalt, rich in iron (and thus in olivine crystals). The vent
that discharged the lava can be seen as a cavity at the foot of a 120-m-
tall cinder cone, in the southwestern corner of the crater. Named Pu'u
Pu'ai ("Fountain Hill"), the cone was formed by vigorous lava fountaining as abundant gas propelled the magma skyward – a behavior
typical of hawaiian eruptions. One fountain reached a record height
of 550 m.

Outside the summit caldera, roads and hiking trails explore the
active rift to the east. The Chain of Craters road follows the upper
section of the rift and passes several craters: Lua Manu and its lava
tree molds, Hiiaka crater, and the triple pit of Pauahi crater.

At the level of the Mauna Ulu lava shield, a side road and
parking area mark the start of the Napau trail, which proceeds farther
along the rift zone. Besides the Mauna Ulu shield, points of interest
on this trail include a perched lava pond (solidified), a variety of lava
flows and two large craters (Makaopuhi and Napau). The round trip

along the trail takes the better part of a day and should be timed so as to return to the parking lot before dusk (take along food, water and flashlights).

Below the rift, the Chain of Craters road descends to the coast. Notice the steep scarps (Holei Pali, Kealakomo Pali) on the way down, caused by the sliding of the entire mountain face along major faults. The road ends at the recent lava field, fed from Pu'u O'o volcano (more than 15 km up slope) via lava tubes. If the eruption is still ongoing, a ranger's booth gives out information on its status, as well as safety recommendations. Beware that the lava crust is sharp as glass, and creeping lava is dangerous (it can surround you, as well as set off methane explosions in overgrown areas). The shelves where the lava tubes reach the ocean are extremely unstable and can break off and slide into the sea with no warning. As in all volcanic areas, warm clothing and windbreakers, good hiking boots, gloves, flashlights, energy food and large amounts of water are mandatory.

Etna   (Sicily)

Latitude:	37° 44′ N
Longitude:	15° 00′ E
Dimensions:	47 km × 38 km (1250 km^2)
Altitude:	3300 m
Volume:	350 km^3
Slope:	3° to 35°
Age:	<500 000 years
	currently active
Caldera	
Diameter:	2.5 km
Depth:	filled with lava

FIGURE 2.10 Mount Etna, photographed by the *Landsat 5* satellite. Recent lava fiows are dark. Credit: *Landsat*/ESRIN.

Towering over the island of Sicily, Mount Etna is one of the most active and best-studied volcanoes on Earth. It is also one of the most diverse in terms of tectonic setting, morphology, magma evolution and eruptive behavior.

Mount Etna is built atop continental crust (sandstones, limestones and clay) at the intersection of three major fault systems (NNE–SSW, NNW–SSE and E–W). Although an island arc of volcanoes (the Eolian islands) lies 100 km to the north, marking the convergence of the European and African plates, Etna is not a subduction volcano. It lies at the corner of the African plate front, where compression is expected to close fissures and prevents magma from reaching the surface. Moreover, the chemistry of Etnean lavas is distinctly alkaline, which is not typical of subduction zones.

Mount Etna is believed instead to be a hot spot volcano, fed by a deep mantle source that happens to lie in a subduction setting. The rising diapir of mantle material lifts and stretches the crust, opening magma pathways to the surface.

Seismic studies and mineral compositions indicate that the diapir stalls and spreads at a depth of 20 to 30 km, at the mantle–crust boundary. From there, offshoots of magma reach the surface to build the volcano, either directly or after pausing at 2 km depth in an upper chamber where the magma decants and changes slightly in composition. Typical lavas at Etna are *trachybasalts* – basalts that are rich in sodium and potassium.

Over 200 000 years, this hot spot activity has created a mountain of lava 3350 m tall, broadly elongated in a north–south direction (48 km ×38 km). From sea level, the volcano rises as a gentle shield (3° slope) up to an elevation of 500 m, then steepens noticeably (10° slope) up to an altitude of 1800 m. Atop this basal shield, stubbier lava flows build a steeper cone (20° slope) that rises to 2900 m, where a terrace defines the emplacement of a 2500-m-wide filled-in caldera: Piano del Lago.

On this plateau, a terminal array of steep cones (30° to 35° slopes) constitute the summit craters – four at the time of this writing. They are frequently active, spewing gas and ash, strombolian projections, or occasional lava fountains and lava flows.

Another characteristic of Mount Etna, which stands out on satellite imagery, is the number of "parasitic" cones that pepper the edifice. These small cones mark the emplacement where streams of magma rose directly from the mantle reservoir to feed short-lived eruptions.

Parasitic cones are typically 50 m to 250 m high. They are not distributed randomly but cluster along three major rift zones: north, south and west.

On the eastern flank of Mount Etna, instead of a classic rift zone, there is a broad depression (8 km × 5 km) opening up on the coast, named Valle del Bove. It marks the collapse of the eastern flank of the volcano into the sea.

Many lava flows are erupted into the Valle del Bove from fissures in its headwall, such as the recent 1992–93 flow. Other histori-

FIGURE 2.11 Shuttle radar image of Mount Etna, showing the summit crater pits and the horseshoe-shaped sector collapse (Valle del Bove) bottom. North is right. Credit: NASA/JPL/JSC.

cal eruptions have surged from fissures lower down on the volcano's flanks, like the 1669 lava flow that broke out at an elevation of 1000 meters and reached the city of Catania. In Etna's more distant past, thousands of years ago, eruptions at the summit were particularly violent and led to large pyroclastic flows. All this variety and history comes to life in a field trip.

Itinerary

Etna towers over the city of Catania on the northeastern coast of Sicily. The best time to visit is May through October. During the winter months, the cold weather and the snow cover at the top of the mountain severely limit its access. In all seasons one must exert caution, dress warmly and carry food and water. Because of altitude (3300 m), the upper flanks of Mount Etna can be cold, even in summertime. They are bombarded by UV rays and subject to dense fogs, storms and abrupt changes in weather. Because of sporadic, unpredictable eruptions, climbing the summit cones is authorized only to experienced volcanologists. Guides periodically take tourists to the base of the cones, when conditions permit. A helmet, gas mask and safety goggles (to protect the eyes from swirling ash) are strongly recommended.

Let us begin our tour of Mount Etna at sea level, just north of Catania. Tectonic uplift exposes the bottom layers of the volcano, revealed in cross-section by the erosion of the sea. These consist of pre-Etnean lavas that were emplaced 500000 to 250000 years ago, when the rising plume started to uplift and fracture the crust. These lavas are a cross between tholeiitic basalt and alkali basalt – the signature of incipient hot spot magmatism. These flows were erupted underwater, as can best be seen at Aci Castello, 4 km north of Catania. In this coastal village, the medieval castle was built on a platform of *pillow lavas* and *hyaloclastites* (submarine lava and ash). The sea-cut bench around the castle is a remarkable place to view these formations.

We leave Aci Castello to drive up the volcano proper, which

FIGURE 2.12 Climbing towards the Piano del Lago summit plateau, with a lava flow to the left and a curtain of ash raining down from the sky. Credit: photo by the author.

began to build up above sea level some 200000 years ago. By then, the deep magma reservoir was churning out alkali basalt, most of which is now buried under more recent lava flows. We head north up the shield, towards the town of Nicolosi. Past the urban sprawl of Catania, the road winds through orchards and vineyards, which exploit the mineral-rich soil.

The lower shield of Etna is a pile-up of basaltic lava flows. At the rate at which Etna erupts (one lava episode every 5 years on average), nearly all flows visible are less than 1000 years old. At Nicolosi begins a historic flow that emerged from a fissure in 1669. This is a typical "excentric" eruption: the magma came up directly from the mantle reservoir, 15 km south of the central chimney. Its high gas content created vigorous lava fountains at the vent, creating two cinder cones, 250m tall: the Monti Rossi. During this voluminous eruption that breached the walls of Catania 12 km to the south, approximately $1.5 \, km^3$ of lava and ash were emitted over a 4-month period.

Driving out of Nicolosi and continuing uphill, one crosses younger flows that were erupted at or near the summit craters. Their source magma spent months or years decanting in a crustal magma chamber, 2 km below the surface, before erupting. These lavas are slightly evolved and fall in the *trachybasalt* category. They are rich in pearly-white feldspar crystals, visible to the naked eye.

As one drives up the volcano, the slope reaches 20 degrees. Above 1200m, the vegetation switches to oak, chestnut and pine trees. Here, recent lava flows cover a greater percentage of the land, burying older ones and steepening the upper flanks. There is no vegetation on these great tracts of rubble and we can pick out channels and levees. After crossing the 1892 and 1910 flows, the road climbs the fresh 1983 flow up to the Cantoniera parking lot and hotels, at 1900m elevation. This is an excellent place to spend the night and get accustomed to the altitude before exploring the summit.

Many side trips are worth taking from here, starting with a spectacular array of cinder cones towering above the parking lot: the 1892

FIGURE 2.13 The southeastern cone at the summit of Mount Etna.
Note the cloud ring expelled by one of the northern pits. Credit: photo
by the author.

Monti Silvestri, prolonged uphill by the recent 2002 cones. Another
side trip is the Valle del Bove. Continuing the road that proceeds from
the Cantoniera down the eastern flank of the volcano towards the
town of Zafferana Etnea, one reaches after about one kilometer a turn
off to the left with a gate and a hiking path. After a kilometer or so
along this path, it is possible to cut left (north) for about half a kilometer across the grassy plateau, and reach the cliff of the Valle del
Bove – a 5-km-wide amphitheater. The steep cliffs of this sector collapse are 600 to 1100 m high in places. A few paths lead down to the
valley floor, but remember, if you descend, you will ultimately have
to climb back out! Always carry a lot of water, a survival blanket and
energy food.

The ash layers visible in some places along the walls are indica-
tive of explosive eruptions, due to the differentiation of magma in the
magma chamber, creating volatile-rich trachyandesite and trachyte.
These pyroclastic eruptions were frequent 100 000 years to 50 000

years ago and were followed by the opening up of the Valle del Bove, which now displays these layers in cross-section. The erosion of the valley walls shows spectacular vertical *dikes* that jut out of the softer ash and lava like giant ribs.

Recent lava flows cascade down the amphitheater's headwall and cover the valley floor. The 1991–93 eruption, in particular, covered 8 km^2 of land and represents 0.25 km^3 of lava – the most voluminous eruption at Etna over the past 300 years. The continual loading of the valley contributes to the seaward slipping of that entire sector of the volcano, measured by radar interferometry (*ERS* satellite) to average 1 cm per year.

To visit the dangerous and unpredictable summit craters, one should hire an Etna guide at the Cantoniera cable car terminal. The cable car (when running) drops off passengers at the 2500 m level. A four-wheel drive then takes visitors to the Piano del Lago, a broad plateau at 2900 m elevation which represents the filled-in caldera of the 122 BC eruption. The steep slope that leads to the plateau is covered with *aa* flows lined with levees – a railroad track pattern that is spectacular from the air and resembles motifs on Olympus Mons and on other Martian volcanoes. Indeed, geologists rely on this similarity to speculate that some flows on the red planet are also made of trachybasalt.

The terminal cones that tower over the Piano del Lago are steep (30° to 35°) and unstable. Progress is difficult and violent eruptions can shower hikers with a deadly barrage of hot scoria. Unless one plans a thoughtful, purposeful climb with a guide (with helmet, gloves, gas mask and goggles), one is best advised to watch the scene from below and from a respectable distance. The craters are a showcase of eruption styles, alternating plumes of ash, hawaiian lava fountains, strombolian projections, and lava breakouts.

Ol Doinyo Lengai   (Tanzania)

Latitude:	2° 45′ S
Longitude:	35° 54′ E
Dimensions:	13 km × 9 km
Altitude:	2886 m
Relief:	2000 m
Slope:	up to 42°
Age:	<370 000 years
Crater:	300 m diameter
Depth:	filled with lava

FIGURE 2.14 Spatter cones of carbonatite atop Ol Doinyo Lengai. Credit: photo by B. Demarne (LAVE).

Most lavas erupted on Earth – basalts, andesites and their derivatives – belong to the silicate family. Their crystal lattice is based on silicon and oxygen. In a few rare cases, however, local conditions are such that other elements are concentrated in the crust and dominate the chemistry of the magma. One such melt is *carbonatite*, a lava that combines carbon dioxide and metals like calcium and sodium. Carbonatites are rare: only 350 occurrences are known on Earth and most are intrusive – i.e. pods and dikes of magma that crystallized underground. Some occurrences are explosive events, driven by carbon dioxide gas, which blast funnel-shaped craters at the surface, known as *diatremes*. These deposits are economically important since they often bear diamonds – the high-pressure form of carbon.

True volcanoes with lava flows are the rarest form of carbonatite. Only one such volcano is active today: Ol Doinyo Lengai ("The Mountain of God") in northern Tanzania. Its exotic magma is derived from the last dregs of silica-poor, alkali-rich magma inside an aging continental rift volcano.

Located on the shoulder of Africa's Eastern Rift valley, Ol Doinyo Lengai is a stratovolcano – a cone of lava flows and ash layers that peaks at an altitude of 2886 m. Initially it erupted silica-poor, alkali-rich silicates such as phonolite and *nephelinite*, which make up the bulk of the volcano. Only recently (starting 1250 years ago) did the volcano begin to erupt carbonatites rich in carbon and sodium.

Eruptions at Ol Doinyo Lengai are often explosive, due to the high gas content in the magma (CO_2). When the pressure builds up under a congested vent, the explosion is vulcanian and can lift heavy boulders in the air (as happened in 1917, 1940 and 1966). These outbursts commonly switch to plinian eruptions, propelling dense columns of ash into the atmosphere (up to an altitude of 11 km in 1966). Otherwise, eruptions in the summit caldera are effusive. Lava sputters out of small vents (*hornitos*) – when bubbles of gas spray the lava skyward in strombolian fashion – or build miniature lava flows with as much variety as their silicate counterparts. They show pahoehoe and aa textures, channels and tubes.

Carbonatite lava is unusual in many respects. It is a low-temperature melt at 800–900 K (500–600 °C) that glows dull red at night. In the daytime, hot carbonatite has the appearance of black motor oil and flows with comparable fluidity. Once solidified, its minerals alter readily when exposed to humidity and turn white. In dry weather this whitening takes a few days, but in rainy weather it can occur in a matter of hours. One can easily spot the most recent lava flows: they stand out as black streaks on the older white patches, a delight for photographers!

Itinerary

A number of tour companies organize expeditions to Ol Doinyo Lengai. This solution is preferable to attempting this demanding hike individually. In any event, it is imperative to hire a local guide. One can easily get lost in the grasslands that cover the deeply gullied lower slopes. One should also check with local authorities if any armed bandits are operating in the area and if the volcano is in a dangerous eruptive phase (vulcanian or plinian).

A trip to Tanzania is an occasion to visit other areas of the East Rift valley, such as the wildlife reserves of the Serengeti Plains and Ngorongoro caldera, 50 km to the south. Lake Natron, its waters sat-

FIGURE 2.15 Spatter cones of carbonatite inside the summit caldera. Credit: photo by B. Demarne (LAVE).

urated with sodium, is only 15 km to the north. Throughout the region, there are many maars (steam explosion craters), cinder cones and extinct carbonatite volcanoes (Shombole, 70 km to the north).

Ol Doinyo Lengai is reached by four-wheel drive, west from Arusha (Tanzania) to Makuyuni, and from there towards Mto wa Mbu. Before reaching this village, an earth road branches north to the volcano, towering over the Rift valley.

The stratovolcano is tall (2886 m) and steep (upper slopes of 42°). In the equatorial heat, the climb is particularly demanding and takes from four to six hours. It is wise to start before dawn and to do most of the climbing before the day gets too hot. One should carry a flashlight, eye goggles and gloves, energy food and lots of water, as well as warm clothing for the summit.

On the lower slopes, deep gullies cut into the soft tuff. These ash layers are made up of alkaline silicates (phonolite, nephelinite). Parasitic cones grow out of the volcano flank. Higher up, burnt tree trunks poke through a carbonate-rich ash layer, dating back to the 1917 explosive eruption. A superposed, more recent, black ash layer (1940 eruption) is locally reworked into spectacular dunes.

The upper slopes of Ol Doinyo Lengai are steep and treacherous, made of consolidated ash that breaks into slippery slabs. Past this last obstacle one reaches the summit: a 300-m-wide crater that was blasted open by the 1966 eruption. Carbonatite flows have since filled the crater to the brim and in 1998 began to spill onto the outer slopes.

The filled-in crater is the theater of nearly continuous eruptions: lava spreads and sputters out of conical mounds (hornitos), up to 25 m tall. Clumps of jet-black lava are tossed out of the vents in strombolian fashion and one should be especially cautious in approaching the cones. These are often hollow and can break underfoot, exposing the molten interior. Such summit cones are often referred to as "toy volcanoes" but there is nothing playful about 500 °C lava suddenly rushing forth with the fluidity of motor oil. One should also wear thick clothes and gloves, as well as a helmet and

FIGURE 2.16 Fresh lava flow of carbonatite (black) atop a weathered older flow (white). Credit: photo by B. Demarne (LAVE).

goggles to fend off occasional blobs of lava that can spatter horizontally out of vents and lava ponds.

Ol Doinyo Lengai can also switch gears without warning into a vulcanian or plinian phase of explosive activity, destroying the entire summit as it did in 1917, 1940, 1966, and to a lesser extent in 1993. One should always be prepared to flee, and camping at the summit is not a good idea.

If one follows these elementary safety rules, the activity at the summit is an enchantment for volcano lovers and photographers. Especially if the weather is damp, the black lava turns light brown and then to white in a matter of minutes. Because of the relatively low temperature and small volume of the flows, one can get relatively close to examine their exquisite flow patterns on a miniature scale, such as lava ponding, channels with raised levees and pahoehoe ropy textures. Are these unearthly landscapes a hint of what we might find on the surface of Venus?

Mount Pelée (Martinique)

FIGURE 2.17 Mount Pelée towers over the seafront town of Saint-Pierre. Credit: P. Barois (LAVE).

Latitude:	14° 49′ N
Longitude:	61° 10′ W
Dimensions:	15 km × 11 km
Altitude:	1397 m
Caldera:	2 km diameter (filled-in)
Age:	300 000 years for complex
	40 000 years for current
	volcano
Last eruption:	1929–32

La Montagne Pelée, in the Caribbean island of Martinique, is an example of subduction-style volcanism. It belongs to the island arc of the Lesser Antilles, which formed over a sinking slab of oceanic crust. As in other subduction settings, the sinking of a crustal plate promotes volcanism. The hydrated crust, carrying water-laden sediments, heats up as it descends into the mantle and releases water vapour and carbon dioxide. These volatiles "flush" through the surrounding mantle, lowering its melting point and causing it to fuse. The magma is rich in water and silica and produces a range of lavas at the surface, known as the calc-alkaline suite, which includes alkaline basalt, andesite and dacite.

The arc of the Lesser Antilles stretches along an 850-km front from Saint-Martin in the north to Grenada in the south. The subduction of the western Atlantic below this arc takes place at the slow rate of 2 cm/year and the output of lava is correspondingly low. There are 15 recent volcanoes in the Lesser Antilles, half of which have erupted at least once in historical times (since AD 1500).

Mount Pelée, on the northern tip of the island of Martinique, last erupted in 1929–32, but it is best known for the 1902 eruption that devastated the city of Saint-Pierre and killed its 28 000 inhabitants (there were only two survivors). During this eruption, a plug of viscous, gas-rich magma rose into the summit zone and managed to break through the southern flank. The sudden drop in pressure caused volatiles in the magma to expand violently, triggering a lateral blast of hot gas and rock fragments that rushed down the slope. Such pyroclastic flows are known as nuées ardentes ("fiery clouds") or pelean eruptions. They characterize the activity of Mount Pelée in historical times. Prehistorically, the volcano was also rocked by plinian eruptions that blanketed the area with thick layers of pumice. Activity at Mount Pelée has switched back and forth between plinian and pelean eruptions over the ages. The 2-km-wide summit caldera was formed by a plinian eruption in the thirteenth or seventeenth century.

A particularly large *ignimbrite* eruption occurred 30 000 years ago, connected with the sector collapse of the southwestern part of the volcano into the sea. The horseshoe depression (6 km × 3 km) is a favored pathway for eruptions up to this day.

Itinerary

Mount Pelée is a volcano charged with history and the potential for violence, as attested by the ruins of the 1902 eruption at Saint-Pierre,

FIGURE 2.18 The ruins of Saint-Pierre in the aftermath of the 1902 eruption. A needle of lava is growing out of the summit crater in the background. Credit: courtesy of D. Decobecq (LAVE).

a village 7 km southeast of the summit on the seashore. There is lodging, restaurants and all commodities in Saint-Pierre (7000 inhabitants). A century ago, in 1902, Saint-Pierre was a small city – the economic and cultural capital of the French Caribbean. The awakening of the volcano in April of that year led to hesitation and chaos. After thick ash falls and a deadly mudflow on May 5, a pyroclastic flow (nuée ardente) destroyed the city on May 8, and killed all but two of its 28 000 inhabitants.

A trip to Mount Pelée begins with a visit of Saint-Pierre, which still bears visible marks of the 1902 eruption. A row of ruins (Ruines du Figuier) on the seafront north of town shows how the pyroclastic blast toppled some walls and left others standing. One can also visit the decapitated ruin of the theatre: only the foundations and the front steps remain.

The hike to the summit of La Montagne Pelée is a 4-hour roundtrip from a parking lot (Parking de l'Aileron) on the eastern flank of the volcano, near the village of Morne Rouge (a 20-minute drive from Saint-Pierre). La Montagne Pelée is French for "the bald mountain": there are no trees for shade and one should carry a couple of liters of water per person and avoid the hot hours of the day (an early morning start is recommended).

The trail is well indicated and starts through bushes, with occasional exposures of weathered ash and bombs that date back to the plinian eruption of AD 250. The trail follows a spur (l'Aileron, French for "fin"), which happens to be an eroded dome of andesite, as can be seen in the trail steps carved in the bedrock. The trail then snakes up the terminal slope through grey andesitic scoria to the rim of the caldera.

The caldera is a 2-km-wide depression. Its steep walls drop 100 meters to a moat that encircles two central domes. These are 200-m-tall plugs of viscous lava that grew out of the crater as crumbling masses during the 1929–32 eruption (in the foreground) and the 1902 eruption (to the north).

A trail winds down the caldera wall and steeply up the domes – now a jumble of large blocks of andesite covered with moss. From the

FIGURE 2.19 In the summit caldera, the weathered domes of the 1929–32 (foreground) and 1902 eruptions. Credit: photo by the author.

top of the domes (as well as from the caldera rim), one enjoys a superb view of Saint-Pierre, the seashore and the deep ravines that gouge the volcano's flanks. During pelean eruptions, pyroclastic flows follow these ravines, especially on the southwestern flank where a 3-km-wide horseshoe depression runs to the sea, created by a sector collapse of the volcano.

An instructive hike up this valley begins on the seafront along the main road (D10), a five-minute drive north of Saint-Pierre. The trail head is at a parking lot and bus stop near the entrance of a pumice quarry. The hike is grueling, with little shade along the way (a 5-hour round trip), and one should carry several liters of water per person. After crossing grassland, the trail reaches the western wall of the valley and descends to a small river of deliciously warm water (Rivière Chaude). The cliff shows a spectacular cross-section of the 1902 and 1929–32 mudflows and pyroclastic flows that rushed down the valley. The riverbed is choked with andesite and pumice blocks of all shapes and colors.

Continuing the hike up the river, one reaches a small waterfall. Past the obstacle (a slippery and difficult ledge), the trail reaches the source of the stream that springs from the ground at 30 °C to 40 °C and is surrounded by sulfurous deposits. The volcano's hydrothermal system has caused *phreatic* steam explosions (notably in 1851), but it also supports life. Algae and extremophile bacteria thrive in the warm waters, outside and inside the plumbing system of the volcano.

3   Volcanism on the Moon

When the Solar System was created, some four and a half billion years ago, growing bodies of rock and ice whirled around the Sun, attracting each other and sometimes colliding in the process. Close neighbors became locked in a gravitational ballet, orbiting each other in binaries or even larger systems. Thus were born the planets and their moons.

Most planets, like Jupiter and Saturn, have relatively small moons, compared to their own size. Exceptionally, the Earth has a large satellite, the Moon, which is 3476 km in diameter. Moreover, the Moon orbits at close range, circling the Earth at an average distance of 384 400 km.

Such a tight, balanced system has many attributes. The gravitational interplay is responsible for the tides that sweep our oceans. Tides might have played an important role in the origin of life as water flowed and ebbed along the seashore. The momentum of the orbiting Moon also acts as a flywheel to stabilize the posture of the Earth, preventing its spin axis from wobbling out of control and from disrupting the march of the seasons. And looming so large in our skies, the Moon is a source of wonder that has opened our eyes on the Universe and beckoned us to travel into space.

FIRST IMPRESSIONS

To the naked eye, the Moon is a patchwork of dark and bright areas often pictured as a face: the proverbial Man in the Moon. In January of 1610, Galileo pointed the world's first telescope at the Moon. Its three-fold magnifying power enabled the observer to view the Moon ". . . as having not a smooth and polished surface, but a rough and uneven one, and as is the case with the surface of the Earth, one covered with high elevations and deep hollows . . ." Galileo named the bright patches *terrae* and the dark expanses *maria*, misinterpreting the latter to be oceans. Galileo also described bowl-shaped depressions that peppered the landscape. In 1665, the British chemist Robert Hooke first suggested that they were volcanoes.

As telescopes improved, astronomers mapped an increasing number of these circular craters. Their volcanic origin was bolstered by their similarity to the Phlegrean craters, west of Naples, although much larger. French astronomer Puisieux proposed that lunar craters were collapsed volcanic domes that had vented all their gases. Astronomer Pierre Laplace added that meteorites were volcanic, ejected from the Moon's craters.

FIGURE 3.1 Location of the six *Apollo* landing sites and the three *Luna* sample-return sites on the near side of the Moon. Credit: NASA photo, from H. H. Schmitt, 1991, Evolution of the Moon: *Apollo* model, *American Mineralogist*, **76**, 773–84.

But another origin was proposed for the craters in the late 1800s. Basing their judgement on the newly discovered Meteor Crater in Arizona, a handful of geologists claimed that the Moon's formations were collision scars blasted by giant meteorites. The lack of an atmosphere allowed cosmic debris to strike the lunar surface unimpeded, and the lack of water preserved the craters from being eroded over the aeons.

The origin of lunar craters remained a hot issue throughout the first half of the twentieth century. Volcano supporters argued that bright rays fanning out of some craters were streaks of volcanic ash, similar to those around volcanoes like Mount Aso in Japan. Astronomers also reported flashes of light and red clouds over craters Alphonsus and Aristarchus. Were these transient phenomena volcanic eruptions, meteorite impacts or optical illusions that tricked exhausted observers?

Less controversial were the smooth, dark maria that were not seas, as Galileo had surmised, nor sedimentary deposits shed off from the highlands. The spectral analysis of sunlight reflected off their surface pointed to iron-rich basalt. The lunar "seas" were thus giant lava fields, comparable to flood basalts on Earth. No obvious volcanoes were associated with these gigantic flow fields, which did not come as a surprise. Flood basalts erupt at such high discharge rates and low viscosities that they fail to build shields and cones around their vents.

CLOSE-UP VIEWS OF THE MOON

With the onset of space travel in the late fifties, the Moon became the target of a grand political race that pitched the United States against the Soviet Union. Photographs of the Moon taken at dedicated observatories were assembled into detailed base maps. Mission planners and geologists selected sites for future robotic and manned landings.

Robots came first. After scoring the first hit of a man-made object on the Moon (*Luna 2* in December of 1959), the Soviet Union achieved the first soft landing on the surface with *Luna 9*, on January 31, 1966. It beamed back black and white images of a granular soil.

More spectacular were the American *Surveyor* probes that scored five successful landings in the Ocean of Storms (*Surveyors 1 and 3*), the Sea of Tranquillity (*Surveyor 5*), and the vicinity of craters Copernicus and Tycho (*Surveyors 6 and 7*). Their cameras revealed breathtaking landscapes, strewn with angular rocks ranging from a few centimeters to several meters in size (see Fig. 3.2).

The *Surveyors* also performed the first chemical analyses of lunar soil, with an alpha ray spectrometer mounted on a robotic arm. The data beamed back to Earth indicated that lunar minerals, like their terrestrial counterparts, were composed essentially of oxygen and silicon, with abundant aluminum, calcium, iron and magnesium, and relatively low concentrations of the alkali metals sodium and potassium. Basaltic lava was consistent with these measurements.

Meanwhile, orbiting probes transmitted thousands of low altitude photographs of the lunar surface, showing details less than ten meters across. These *Lunar Orbiters* showed the Moon to be markedly asymmetrical. The far side was battered by impact craters and impact basins, but lacked the vast, smooth maria that were typical of the near side. Detailed imagery of the maria showed overlapping patches of lava, flow fronts and wrinkle ridges caused by the buckling of the surface. Sinuous rilles that snaked across the landscape resembled dried-up river beds, but were more likely giant lava channels.

Conspicuously lacking, however, were true volcanoes. There were no lava shields, no cones with summit craters, no collapsed calderas and no puffed-up domes. The only landforms that showed a liking to volcanoes were clusters of hills on the periphery of some maria, including the Harbinger domes at the foot of the Aristarchus plateau, and the Marius Hills farther south. The Marius Hills, in fact, raised enough interest to make it into the final list of candidate landing sites for the *Apollo* missions.

APOLLO 11: TOUCHDOWN IN A LAVA FIELD

As it turned out, the target of the first manned landing on the Moon was chosen essentially for reasons of safety, rather than for scientific

FIGURE 3.2 One of the first ground pictures of the Moon, taken by
Surveyor 7 in January 1968, north of the impact crater Tycho. The
horizon is one kilometer away. The probe analyzed the soil and found it
to be richer in aluminum and sodium than the mare sites. Credit:
NASA.

interest. The primary target for *Apollo 11* was a smooth-looking patch
of sparsely cratered plains, on the western edge of the Sea of Tranquillity
(see Fig. 3.1).

On July 20, 1969, Neil Armstrong and Buzz Aldrin successfully landed their lunar module *Eagle* on target, and looked out on a dusty plain, peppered with rock fragments. The thick soil that blanketed the site resulted from scores of meteorites pounding the lava flows over millions of years. When Neil Armstrong took his first step on the Moon, he described the surface as "fine and powdery. I can pick it up loosely with my toe. It adheres in fine layers like powdered charcoal to the sole and sides of my boots . . ."

Buzz Aldrin, stepping on the surface in turn, was fascinated by the pristine aspect of lunar rocks and the shiny specks that reminded him of flakes of mica. As it turned out, the sparkling was caused not by mica – mica is a water-bearing mineral and there is no water on the Moon – but by crystals of pyroxene and feldspar, and by beads of glass.

Armstrong guessed the rocks to be basalts. Many were vesicular, riddled with bubbles that formed in the magma as the molten rock cooled and degassed on the lunar surface. During their two-hour moon walk, the astronauts bagged 21 kg of rock and soil samples, all collected within 50 meters of the lunar module – the area of a football field. Despite their small playground, they brought back to Earth a representative and surprisingly varied assortment of lunar rocks that fell into two broad categories: chunks of lava from the local lava flows, and impact *breccia* – rocks mixed and fused together by meteorite impacts, which flew in from distant sites.

Dispatched to geologists around the world, the *Apollo 11* lavas were confirmed to be basalts, characterized by the minerals feldspar and pyroxene, occasional olivine, and a sprinkling of iron and titanium oxides. Chemically speaking, the lavas as a whole contained 40% silicon, 18% iron, 12% calcium, 10% aluminum and 10% titanium oxides. Relative to terrestrial basalt, one striking difference was the abundance of titanium, a refractory metal with a high melting point (basalts on Earth contain less than 1% titanium oxide). Another difference was the dearth of volatile elements – elements that melt at modest temperatures – such as sodium (0.4%) and potassium (0.01%). In particular, there was no hydrogen and thus no water-bearing minerals in the lunar rocks.

Some process had enriched the Moon in refractory metals and depleted it in volatile elements, a finding that scientists would need to address when looking to explain the creation and evolution of our satellite.

In the lunar lavas, radioactive elements like potassium, uranium and rubidium allowed geochemists to date the rocks by the method of radiochronology. The decay of such elements takes place at a given rate, and the proportions of "parent" and "daughter" elements (potassium decays to argon, for example) gives an estimate of the elapsed time since the rock's formation, i.e. its age.

Performed on the *Apollo 11* samples, this technique yielded ages of 3.7 to 3.8 billion years for the high-titanium basalts, and 3.6 billion years for a slightly younger type of basalt containing less titanium. Hence, the samples date back to the first quarter of lunar history (the Moon, like all planetary bodies, is 4.6 billion years old). As expected, the Moon was turning out to be a showcase of the earliest forms of volcanism in the Solar System – a priceless planetary museum.

APOLLO 12: THE OCEAN OF STORMS

Four months after the triumphant mission of *Apollo 11*, a second crew headed out for the Moon's volcanic maria. Pete Conrad and Alan Bean made their landfall on November 19, 1969, in the southern stretches of the Ocean of Storms, 300 km south of Copernicus crater. In a feat of precision landing, Conrad dropped the lunar module a mere 200 meters from an unmanned *Surveyor* probe that had landed there two years prior.

Apollo 11 had proven the lunar maria to be giant sheets of basalt, that had flooded the impact basins on the Moon's near side. But had the lavas welled up from the depths of the Moon in response to the large impacts that carved out the basins, or had they erupted on their own, independently of the basin-forming events?

Apollo 12 would help clarify the issue. The landing site was located in a rolling lava plain, battered by impact craters the size of

FIGURE 3.3 *Apollo 12* landed on the lava flows of the Ocean of Storms, a few hundred meters from the automatic *Surveyor 3* probe that landed two years prior. Alan Bean inspects the spacecraft, photographed by Pete Conrad. The lunar module is visible on the horizon. Credit: NASA, archives of Goddard Space Flight Center (NSSDC).

football fields. From orbit, the site showed at least two separate lava flows with slightly different densities of impact craters, suggesting different ages. *Apollo 12* landed on the oldest flow. The samples collected by the astronauts turned out to be about 3.3 billion years old, half a billion years younger than the rocks brought back by *Apollo 11*. The other unsampled flows in the vicinity of the landing site were probably as young as 2.7 billion years.

These dates confirmed that lunar lavas were very ancient, but still spanned a wide interval of time. Most had erupted much later than the epoch when the basins were formed. This showed that volcanic flows erupted on their own, from the deep mantle, and that melting was not simply triggered at the surface by the large impacts. However, impact basins certainly facilitated the ascent of magma, since their fractures provided easy pathways to the surface.

Besides their spread in age, lunar basalts also spanned a range of chemical compositions. Whereas *Apollo 11* basalts were exceptionally rich in titanium oxide, those collected by *Apollo 12* contained much less of this exotic element (3%, down from 10%). They came in three varieties: quartz basalts, containing large pyroxene crystals visible to the naked eye; olivine basalts, sparkling with crystals of the green mineral; and one peculiar, very aluminous specimen.

Besides collecting samples, Conrad and Bean set up an array of scientific instruments, including a seismometer. By measuring seismic waves travelling through the crust, as well as by studying the topography of the site from orbit, geologists estimated that the lava flows were 200 meters thick. Elsewhere in the Ocean of Storms, some flows are over 1000 m thick, and in the deeper basins of the Moon they reach an aggregate thickness of 6000 m. These figures are comparable to those of the great lava fields on Earth – the flood basalts of India, Siberia or the Columbia River Plateau. Like their earthly counterparts, lunar flows welled up as very fluid magma, and erupted at very high discharge rates. They covered large areas, buried their own vents and left little clues as to where they originated and how they erupted. Did the magma jet upward in fiery fountains high above the lunar surface? Were there occasional blow-outs of hot ash, rolling down the lunar slopes?

The next two *Apollo* missions did little to answer these questions. *Apollo 13* aborted (a fuel cell explosion) on the way to the Moon and failed to land on the surface, but the crew returned to Earth safe and sound. Alan Shepard and Edgar Mitchell, on *Apollo 14*, explored the Fra Mauro hills of the Ocean of Storms, which turned out to be piles of impact breccia with few volcanic specimens among the samples. Fortunately, much insight into eruptions on the Moon was provided by *Apollo 15*, which set out in July of 1971 for the eastern edge of the Sea of Rains and a meandering channel known as Hadley Rille.

APOLLO 15: EXCURSION TO HADLEY RILLE

Hadley is one of several hundred sinuous rilles that cross the lunar maria. Most rilles run between 10 and 100 km in length and span up to 3000 m in width. They resemble rivers, but there is no water on the Moon. Before the flight, the ruling hypothesis was that lunar rilles were carved by turbulent outpourings of lava that followed fault lines down to the plains. Exploring a rille was thus an excellent opportunity to study volcanism on the Moon. The target of *Apollo 15*, Hadley Rille emerges from an arcuate cleft in the Apennine mountains of the Moon and descends to the floor of the Sea of Rains (mare Imbrium), where it runs northward along the edge of the basin. The rille is

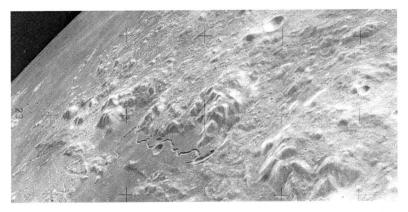

FIGURE 3.4 Hadley Rille, the target of *Apollo 15*, starts in a cleft in the Apennine Mountains and winds its way along the northern margin of the Swamp of Decay, on the eastern margin of the Sea of Rains. Credit: NASA, courtesy of the National Space Science Data Center, through the World Data Center-A for Rockets and Satellites.

100 km long, averages 1200 m in width and reaches a depth of 300 to 400 m along much of its course.

On July 31, 1971, astronauts Dave Scott and Jim Irwin flew their lunar module over the Apennine mountains and down onto the lava plains, landing one kilometer east of Hadley Rille's right bank (see Fig. 4.3). For the first time in the *Apollo* program, they brought along an electric-powered lunar rover, folded up against their spacecraft. Over the course of three outings – each lasting six to seven hours – the astronauts drove 28 km around the landing site, making stops at ten different locations to take photographs, drill cores and collect a record heist of 77 kg of rocks and soil.

On their first outing, Scott and Irwin drove due south, to the base of the Apennine mountains, where Hadley Rille hugs the foothills, then takes a sharp turn and strikes northwest across the lava plains. From their vantage point they caught a good view of the rille, took samples and a series of spectacular photographs (see Fig. 3.5) but had to press on. It wasn't until their third outing that the mission plan called for a detailed study of the rille, farther north, as well as a look at a cluster of craters that might be volcanic vents.

On Tuesday, August 2, the astronauts finally headed out to explore the rille. Reaching their objective, known as the "terrace," they parked the rover on the rim of the channel and looked across at the opposite bank. The rille wall showed layers of basalt poking out of the talus like bony ribs. Scott and Irwin managed to lope a few steps down slope and collect rock samples from broken slabs of tan-coloured lava, before being called back by the anxious officials at mission control. Time was running short, and to the astronauts' disappointment, the last stop at the northern craters was cancelled. The mystery of their origin – meteoritic or volcanic – would be left to future explorers. In the next chapter, we devote our first lunar field-trip to this wonderful and historic site (pages 97–103).

Hadley Rille brought a new perspective to the study of volcanic flows and channels. On Earth, there are no features of such dramatic size, although small tubes and open channels are plentiful on the slopes of basaltic shields, as in Hawaii and La Réunion islands. Such channels carry smooth lava flows that leave a chilled lining of basalt on their floor and sides, raise their beds and build levees that often roof over to create tunnels. Such flows are "constructive" in the sense that they create a positive relief.

On the other hand, when the discharge of magma is enormous, as it was at Hadley Rille, the flow is no longer peaceful and stratified with the hottest magma in the centre and the cooler material on the outside. Instead, the flow is turbulent, rolling and mixing. Rather than lining the floor of the channel with layers of chilled basalt, it cuts down through the bedrock, melting and plucking away an ever-deepening channel bed. This is all the more likely if the stream of magma finds a weakness in the crust, such as a fault line exposing a soft layer of rock. At Hadley, the rille apparently dug its course along the bounding faults that ring the Imbrium basin.

A detailed history of the site can be reconstructed from the astronauts' observations and the rocks that they brought back to Earth. For beginners, Scott and Irwin picked up a chunk of basement rock, shed from the uplifted mountain range that dates back to the

FIGURE 3.5 Jim Irwin and the lunar rover on the rim of Hadley Rille. The view is to the northwest. In the foreground, the rille is nearly 1500 m wide and 300 m deep. In the distance (10 km away), the rille takes a bend to the left. Credit: NASA, courtesy of the National Space Science Data Center, through the World Data Center-A for Rockets and Satellites.

earliest days of the Moon. Nicknamed "the Genesis rock," this sample of *anorthosite* (a rock made up mostly of feldspar crystals) is 4.5 billion years old.

The Imbrium impact that uplifted the Hadley mountain range is dated at 3.85 billion years, according to breccia collected at the site. Volcanic activity began at that time, as aluminous basalt welled up through the fractured crust in the aftermath of the impact. Named the Apennine Bench Formation, this early basalt is still visible along the basin rim.

The majority of lava flows that fill the Imbrium basin are much younger. Those that flowed down Hadley Rille are 3.3 to 3.4 billion years old. They are predominantly pyroxene basalts, with an olivine variety forming a darker layer at the top of the rille. Apparently, this olivine basalt was the last flow to race down the channel.

Besides hard rock, the astronauts also collected spadefuls of a sparkling emerald-green soil. Back on Earth, viewed under the microscope, the soil was resolved into tiny beads of volcanic glass, tinted by various metals (the green tint was due to magnesium). These droplets of magma were formed in a lava fountain that accompanied the mare eruptions, over three billion years ago. The glass beads were later scattered around the site by the relentless pounding of meteorite impacts. As for the eruptive vents themselves, they remained to be discovered, if anything was left of them. One suspect was the cluster of craters at North Complex (that the astronauts were unable to visit for lack of time), although an impact origin was considered more likely. Other candidate volcanoes were identified on orbital imagery across the maria, and made it into the list of prospective *Apollo* landing sites.

APOLLO 17: A GEOLOGIST ON THE MOON

In the Spring of 1972, *Apollo 16* set off to explore the Descartes region of the lunar highlands. Its target was a patch of smooth terrain amidst a jumble of hills and craters, which looked suspiciously like an ash fall deposit or some other form of highland volcanism. Before the mission, some geologists had gone so far as to map out lava flows and cinder cones in the area. But when John Young and Charlie Duke explored the landing site, they found only impact breccia – "puddings" of shattered rock, tossed about the Moon by impacts far and wide – but no evidence of highland volcanism.

Apollo 17, the last flight to the Moon, would provide one last chance to look for "fresh" volcanoes, while also providing access to the oldest lunar crust. This dual objective was filled by the valley of Taurus-Littrow, a young-looking lava plain nested between uplifted blocks of an ancient mountain range, on the eastern margin of the Sea of Serenity (see Figs. 4.6 and 4.7).

During the *Apollo 15* mission, astronaut Al Worden had conducted a thorough survey of the area from lunar orbit with a telephoto camera. Of great interest in the photographs was a prominent landslide scarring the southern mountain face, which would give the astronauts the opportunity to sample ancient material shed off the slopes. There was also a sprinkling of dark craters on the valley floor that looked like recent volcanic vents, surrounded by their ash fallout.

Apollo 17 was also the occasion of a great premiere. Breaking the monopoly of professional pilots, a civilian scientist would finally fly to the Moon: geologist Harrison "Jack" Schmitt. The Harvard graduate had joined the Astronaut Corps in 1965, and spent the next seven years in training, learning how to fly fighter jets – he had never flown a plane before – and running the full gamut of mission rehearsals. His patience and dedication were rewarded. On December

FIGURE 3.6 Astronaut Jack Schmitt investigates a boulder at the foot of North Massif, on the Taurus-Littrow site. This boulder was part of the crustal layers uplifted by the Serenitatis impact and contains impact breccia that resembles volcanic lava. Credit: NASA.

6, 1972, Jack Schmitt took his berth alongside Gene Cernan and Ron Evans aboard the *Apollo 17* spacecraft, acting as a flag bearer for the entire science community, and blasted off to the Moon.

On December 11, Cernan and Schmitt nailed a pinpoint landing in the Taurus-Littrow valley, less than 200 m from their target. Focusing on his pilot's job – reading out altitude and velocity numbers to his commander – Schmitt was not able to look at the lunar surface on the way down. But once on the ground, he would more than make up for the lost opportunity, although working in a spacesuit came with its share of frustrations.

The view at Taurus-Littrow was breathtaking. The astronauts faced a flat boulder-strewn valley, 50 km long and 7 km wide, boxed in by lofty mountains that tower 2000 m above its floor. Large boulders had rolled down the slopes, leaving bright grooves in their wake. On their first outing, Cernan and Schmitt installed a science station and had little time to sample rocks, let alone gaze at the landscape. In fact, Schmitt later noted in his mission report that unlike field trips on Earth, lunar traverses were marred by the pressing quest for speed and efficiency, which prevented him from consciously memorizing his visual impressions. Only during debriefing sessions on Earth was the astronaut able to retrieve some of the observations stored in his subconscious.

During their first quick pick of rocks near the lunar module, Cernan and Schmitt netted basalts very similar to the *Apollo 11* samples. They were of the same age (3.7 billion years) and displayed the same high content of titanium oxide (10%). This did not come as much of a surprise since the *Apollo 17* landing site lay on the edge of the Sea of Serenity, close to where it merged with *Apollo 11*'s Sea of Tranquillity.

The astronauts' second outing, set to last seven hours, was dedicated to an extensive survey of the valley. Four major stops were scheduled on the 18-km loop. The first stop was the avalanche apron at the foot of South Massif. The avalanche boulders were made of very old breccia, uplifted into a mountain range by the blast that created

FIGURE 3.7 Astronaut Jack Schmitt, on the rim of Shorty crater, packs away the core of orange soil that he drilled on the site. The 110-m-wide impact crater lies out of frame to the right. For color images of the crater and the orange soil, see color Plates IX and X. Credit: NASA, photo by Gene Cernan.

the Serenity basin. The boulders that rolled down the mountain front thus offered a glimpse into the early days of lunar history, and the explorers spent a full hour combing over the site.

Struggling to remain on schedule, Cernan and Schmitt drove back down into the valley, towards a couple of small dark craters that poked through the light-coloured landslide material. These were the heralded "young volcanoes" that geologists had spotted on the orbital imagery, surrounded by mantles of dark ash. Now, "ground truth" was about to deliver a verdict.

LAVA FOUNTAINS ON THE MOON

The astronauts parked their rover on the edge of Shorty crater, one of two candidate volcanoes. To Schmitt, the 110-m-wide bowl looked like one more impact crater, with its characteristic ejecta blanket of shattered rock. But as Schmitt moved in to take a series of photographs, he stopped suddenly in his tracks, staring at the ground.

"Wait a minute!" he exclaimed, "the soil is orange!"

Schmitt lifted his gold-plated visor to make sure the orange color was real, and not some reflection cast by the nearby rover. The orange tint did not go away. Cernan hurried over. Kicking up the soil with his boot, Schmitt uncovered a streak of bright orange soil, stretching four to five meters along the crater rim.

"If I ever saw a classic alteration halo around a volcanic crater, this is it!" ventured Schmitt, hinting that the soil had been oxidised by water-rich gases. Such a discovery would topple the belief that the Moon was waterless. Moreover, the geologist guessed that the fresh-looking deposit was only a few million years old, reviving the hopes of recent volcanism on the Moon.

In great haste, with only half an hour to spend at the site, the astronauts gathered spadefuls of orange soil and drove a core through the top layers. The orange deposit was only 20 cm deep, underlain by purple and black soil. Its edges were sharp and vertical, with the colour grading from dark crimson at the centre to yellowish orange on the margins. The colour pattern was reminiscent of the oxidation of soil along a volcanic fissure zone, as the astronauts had witnessed many times on their field trips to Hawaii and other volcanoes on Earth. But why did the orange layer end abruptly, less than a foot below the surface?

There was no time to push the investigation any further. Air reserves were getting low and Houston was calling for a wrap. The astronauts boarded their rover and drove back to the lunar module, leaving Shorty crater and its mysteries behind.

The discovery of the orange soil was the highlight of the mission. Back on Earth, the samples were unpacked with great anticipation and run through the gamut of preliminary tests. The soil turned out to be composed of microscopic droplets of volcanic glass, the size of pinheads. Their chemical make-up was basaltic, but there was no trace of hydrothermal oxidation. In fact, the beads were

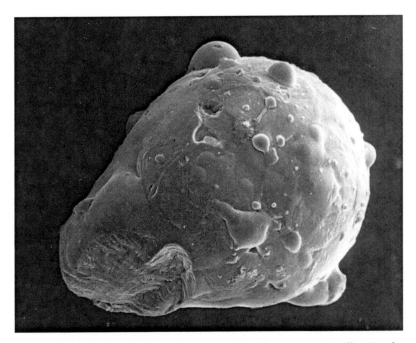

FIGURE 3.8 The orange glass retrieved at Shorty crater (*Apollo 17*) and the emerald green glass retrieved at Hadley Rille (*Apollo 15*) are droplets of magma that were shot up into the vacuum of space and took on spherical shapes. Credit: NASA/Johnson Space Center, courtesy of Dr. James L. Gooding.

colored orange because they were glassy and contained titanium. The purple beads were enriched in iron. They shared the same origin as the green glass of *Apollo 15* and resulted from a spray of lava droplets out of an erupting vent, spun and chilled into glass. On Earth, hawaiian lava fountains create similar fall-out, albeit larger in size, called *scoria*. Some droplets are strung out during their flight into glassy filaments known as *Pele's hair*, in tribute to the Hawaiian goddess of volcanoes. On the airless Moon, the beads were mostly spherical.

But the biggest surprise – and a disappointing one to many – lay in the age of the *Apollo 17* glass. The verdict was 3.6 billion years, scarcely less than the age of the lava collected elsewhere in the valley (3.7 billion years). The orange soil seemed incredibly young, not because it really was, but because it had been buried by lava flows and protected from alteration. Only recently had the beads been exposed at the surface, so that they were scarcely tarnished by the attack of cosmic rays. In this respect, Jack Schmitt had been right in judging the soil to be only a few million years old: cosmic rays suggest that it was brought to the surface 19 million years ago. The culprit of this exposure was indeed Shorty, not a volcano but yet another impact that broke through the lid of lava and sprayed the glass beads onto the surface.

TREASURES OF *APOLLO*

From the first Moon fall of July 1969 to the farewell mission of December 1972, six *Apollo* crews roamed the lunar surface, traversing a total distance of 100 km and collecting 382 kg of lunar samples (75% rocks, 25% soil). To this one must add the small drill cores returned by the Soviet automatic probes *Luna 16* (1970), *Luna 20* (1972) and *Luna 24* (1976). Although modest in weight (100, 30 and

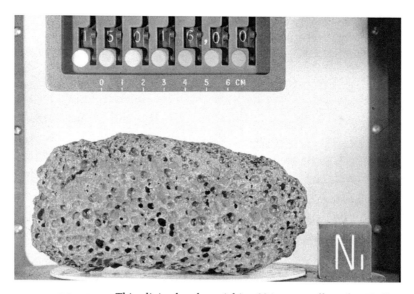

FIGURE 3.9 This olivine basalt, weighing 924 g, was collected at Hadley Rille by the *Apollo 15* astronauts. The abundant vesicles point to degassing of volatiles (probably carbon monoxide) from the magma. Credit: NASA/Johnson Space Center, courtesy of Dr. James L. Gooding.

FIGURE 3.10 Thin section of the *Apollo 15* olivine basalt, viewed under the polarizing microscope. The field of view is 4 mm wide. The mosaic of crystals include large grains of clinopyroxene (center), plagioclase feldspar (striped parallelograms), pearly grains of olivine, and dark specks of glass and metal oxides. Credit: NASA/Johnson Space Center, courtesy of Dr. James L. Gooding.

170 grams respectively), these cores broaden our sampling base from six to nine sites, adding the Sea of Fecundity, the Sea of Crises and the Apollonius highlands to the *Apollo* sextet.

Lunar volcanism is well represented in the sampling. Six sites out of nine are basaltic mare plains. On other sites, samples from older volcanic episodes are excavated by impacts into the crust.

Basalts are the dominant rock family, ranging in age from 3.2 billion years (*Apollo 12*) to 3.8 billion years (*Apollo 11* and *17*), with a few older specimens lodged in the highland breccia. Their chemistry, which owes them the appellation of basalt, is characterized by relatively low silicon and relatively high iron, calcium and magnesium. Some lunar basalts contain unusually large amounts of titanium, a *refractory* metal with a high melting point. In terms of minerals, this chemistry translates into a mosaic of feldspar and pyroxene crystals, with a sprinkle of iron and titanium oxides, and occasional olivine. Because the Moon lacks water, the crystals have not been oxidized or otherwise altered and look remarkably pristine.

The lack of water, on the other hand, severely restricts the range of minerals that can exist in lunar rocks. Whereas over 2000 mineral species are known on Earth (90% of which contain water), only about a hundred have been identified on the Moon. The lunar list is short, but it does include a few original species that owe their existence to the Moon's reducing environment (lacking free oxygen). One such species is a new form of titanium oxide containing iron and magnesium, baptized *armalcolite* in tribute to the *Apollo 11* astronauts Armstrong, Aldrin and Collins. Another is a silicate of iron and titanium named *tranquillityite* in reference to the Sea of Tranquillity. Since their discovery in lunar samples, some of these minerals have also been found on Earth, in rocks formed in reducing environments.

Lunar basalts are further characterized by their extreme dearth in volatile elements. Not only do they lack water, but they are also very poor in sodium, potassium and other elements with low melting points. These chemical traits must be taken into account in models that aim to explain the origin of the Moon.

Besides their distinctive character, lunar basalts vary slightly from site to site. Geologists divide them into several categories, based on their aluminum, potassium or titanium content. The percentage of titanium, in particular, allows us to distinguish three types of lunar basalts: low-titanium basalts (6% to 12% titanium oxide), medium-titanium basalts (2% to 6%) and low-titanium basalts (below 2%). The titanium content appears to reflect the depth of the magma source, with low-titanium magma supplied by the deeper parts of the lunar mantle. Aluminous basalts constitute a special class of lunar lava. *Apollo-14* astronauts Alan Shepard and Edgar Mitchell found fragments of this type at Fra Mauro crater, on the eastern edge of the Ocean of Storms. *Apollo-15* astronauts David Scott and Jim Irwin sampled others in the Apennine foothills, on the edge of the sea of Serenity. Besides being aluminous, these basalts are relatively rich in potassium (K), rare Earth elements (REE) and phosphorous (P), a chemical make-up labeled *KREEP* by the scientists. These "creepy" basalts appear to be the last dregs of a "magma ocean" that once covered the Moon, as we shall later see.

Amateurs of volcanism will be disappointed to learn that the lunar story is limited to basalts. There are no shiny-faced andesites, no peppered trachytes, no colorful rhyolites on the Moon, at least not in detectable quantities. Such "evolved" siliceous lavas are formed on Earth by distillation processes at work in magma chambers close to the surface, mostly in water-rich environments.

But there is no water on the Moon. Moreover, the magma does not appear to stall in crustal reservoirs on its way to the surface, as it so frequently does on Earth. Thus the magma remains essentially basaltic. Besides the evidence on the ground, remote sensing from lunar orbit has recognized only basaltic signatures in the lava plains.

REMOTE-SENSING: THE BIG PICTURE

During the *Apollo* moon landings, the third astronaut in lunar orbit conducted surveys of the terrain below the moving spacecraft, especially Al Worden (*Apollo 15*), Ken Mattingly (*Apollo 16*) and Ron Evans (*Apollo 17*) who operated remote-sensing cameras and detectors. An X-ray fluorescence and gamma ray spectrometer measured the concentration of aluminum, silicon, iron, as well as the radioactive elements uranium, thorium and potassium along the ground track below. However, these surveys were limited to the equatorial region of the Moon, because of the near-equatorial orbit of the *Apollo* spacecraft.

A global survey of the Moon was achieved only much later. The Jupiter-bound *Galileo* spacecraft performed a quick flyby of the Moon in December of 1990, during which it tested its instruments. Several major discoveries were made in the process, including the identification of a giant impact basin near the lunar south pole.

Next, the *Clementine* probe circled the Moon for six weeks in 1994, before flying off to a nearby asteroid. *Lunar Prospector*, another low-cost spacecraft, accomplished a dedicated mission around the Moon for a year and a half in 1998 and 1999. It flew a neutron and gamma ray detector that picked up the signals of elements like iron and titanium, a magnetometer to study relic magnetic fields around impact basins, and a gravitometer to locate accumulations of dense igneous rock below the surface.

From orbit, subtle differences in lava chemistry can be detected. In visible light, they take the form of slight changes of color. Low-

FIGURE 3.11 The Aristarchus plateau, in the Ocean of Storms, is the most outstanding volcanic province on the Moon. A wide lava channel (Schroeter's valley) emerges from Herodotus crater at right of frame. Both craters, however, are of impact origin. Image by *Apollo 15*. Credit: NASA.

titanium basalt displays a reddish hue, medium-titanium basalt peaks in the orange, and high-titanium basalt is bluest. The history of a mare basin can thus be told from its colors. In the Sea of Serenity, for example, eruptions started with blue titanium basalt in the south – those samples collected by *Apollo 17*. Medium-titanium basalt then erupted around the rim of the basin. Lastly, reddish low-titanium basalt flooded the central part of the down-warping basin.

Spectrometers can also recognize the chemical signature of buried lava fields, since their chemical "flavor" is disseminated by impacts into the overlying soil. These "phantom" basalts are known as *cryptomaria*. They have now been detected in the Schiller-Shickard area of the Moon (southwestern limb), in the north polar plains, and in the South Pole-Aitken basin on the far side. Cryptomaria extend the total area of volcanic plains on the Moon from 17% to nearly 20% of its surface.

The vast majority of lava plains are located on the Moon's near side. This can be explained by the layout of the lunar interior. Our satellite is somewhat oblong, rather than perfectly spherical, with its longest axis pointing towards Earth. The tidal pull of our planet has distorted the Moon in such a way that its dense mantle stretches out on the near side, thinning the crust there. When pockets of magma form at the top of the mantle, those on the near side have less of a distance to travel to the surface through thinner crust, and maria are concentrated there.

One can estimate the thickness of the lava pile inside lunar basins by studying the local gravity field. Satellites sense a slight acceleration when they fly over denser parts of the Moon. They show that mass concentrations, known as *mascons*, underlie most impact basins. These gravity anomalies indicate a lava thickness of at least 3500 m in the central area of the Sea of Serenity, and an average of 5000 m in the Sea of Rains (mare Imbrium) with central pockets up to 8000 m deep. In addition, the *Lunar Prospector* probe has identified a dozen mascons on the Moon, where no lava flows are visible. These might indicate regions where dense magma has stalled in the crust without reaching the surface.

Seismic studies of the lunar interior were conducted with

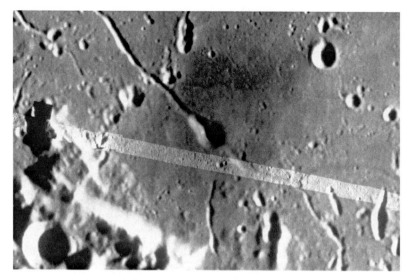

FIGURE 3.12 A dark halo crater on the floor of Schrödinger basin (76° S, 139° E) is interpreted to be a volcanic vent that erupted ash over 3.5 billion years ago. The image was collected by the *Clementine* probe, in the visible and ultraviolet parts of the spectrum. Credit: image processing by the U. S. Geological Survey, Flagstaff, courtesy of LPI (slide set: *Clementine explores the Moon*, compiled by Paul S. Spudis).

instruments deployed on the ground during the *Apollo* missions. According to these seismometers, moon quakes are rare. The total energy released over the course of a year is only *one hundred billionth* of the Earth's seismic output. Moon quakes originate 700 km to 1200 km below the surface and peter out at greater depths, as if the deeper layers of the Moon were partially molten. However discrete, these moon quakes, combined with calculations of the Moon's moment of inertia, allow us to draw a cross-section of our celestial companion.

The crust of the Moon averages 60 km in thickness – somewhat less on the near side – and is underlaid by a rigid mantle of olivine and pyroxene (not unlike the Earth's mantle) down to a depth of 1000 km, where rigidity is lost and the lower mantle is partially molten. Starting roughly at 1400 km depth and down to the center (1734 km), there seems to be a molten iron core. This comes as a surprise, since small bodies like the Moon are not supposed to "melt down" and concentrate their iron into a core. For the Moon to have done so implies a rather catastrophic history.

THE MAGMA OCEAN

The origin of the Moon had long been a puzzle to scientists. Three models were in vogue, prior to the *Apollo* flights. The fission theory held that the Moon spun off from a rapidly rotating Earth, early in the history of our planet. In the co-accretion theory, the Earth and Moon grew together but separately, out of the same cloud of material. The capture theory ventured that the already-formed Moon flew in from elsewhere in the Solar System and was trapped into Earth orbit.

Results from *Apollo* challenged all three views. Lunar rocks

FIGURE 3.13 A large-scale collision between the Earth and a smaller planet might have created the Moon in the early days of the Solar System. Credit: painting by William K. Hartmann.

displayed the same isotope ratios of oxygen (O_{18} to O_{16}) as terrestrial rocks, indicating that their mantles were made up of identical material. However, the Moon contained less volatile elements than the Earth, as if it had lost them to space in some gigantic heating event. Finally, the existence of an iron core at the center of the Moon was at odds with all three models.

A new and unlikely scenario began to emerge from the conflicting data. What if a growing planet, the size of Mars, had collided with the Earth in the early days of the Solar System? Computer simulations showed that the rogue planet would have ripped through the Earth's mantle and mixed most of its iron core up with our own, like two eggs blending their yokes. A plume of hot debris, including some leftover iron, would have streamed off into Earth orbit, rapidly coalescing to form the Moon.

This giant impact theory helped to explain the similar isotope ratios in the Earth and Moon: elements from both bodies were well mixed during the collision event and yielded a common stock. The Moon lacked volatile elements because the hot debris had vented them to space before coalescing to form our satellite. As for the Moon's iron core, it was inherited from the body that collided with Earth – not all of it had blended with the Earth's core.

The creation of the Moon was thus a hot affair. There is ample evidence that its outer layers were thoroughly molten early on, down to a depth of 400 or 500 kilometers – an episode known as the "magma ocean."

One sign of this infernal event is that the Moon's upper mantle is depleted in *incompatible elements* (large ions like europium), whereas these elements are strongly enriched in the overlying crust. Only a massive melting of the upper mantle could explain how such a transfer took place. In a magma ocean, europium and other incompatible elements were scavenged by crystals of plagioclase feldspar that grew and rose buoyantly to the top of the ocean. This feldspar-rich "scum" cooled to form the lunar crust, which is now exposed in the highlands.

FIGURE 3.14 This lunar sample, collected by the *Apollo 17* astronauts from a boulder at the foot of South Massif, is a dunite: an igneous rock made up principally of the mineral olivine. It probably crystallized out of a magma ocean in the early days of lunar history. Credit: NASA/Johnson Space Center, courtesy of Dr. James L. Gooding.

The Moon's magma ocean took on the order of 200 million years to cool and solidify, from 4.55 to 4.35 billion years ago. At the end of this congealing stage, there were only a few pockets of residual magma left in the lower crust. These last pods of decanted magma, rich in potassium, aluminum and phosphorus, finally solidified 4.35 billion years ago, but only for a short time. With low melting points, these pods would be the first to melt back into magma, if the Moon were to heat up again.

There was a good supply of heat at the surface. Falling out of the sky, asteroids and comets continued to beat up and locally melt the lunar crust. They carved out the large basins, spreading sheets of breccia around the Moon. Molten crust trickled and collected in the hollows of some basins, forming smooth plains of impact melt. Some are still visible today inside the "bull's eye" pattern of mare Orientale – the last great impact to take place on the Moon, some 3.8 billion years ago.

MARE VOLCANISM

The thick layers of impact breccia that were sprayed across the Moon acted as an insulating blanket that blocked the flow of heat from the lunar interior (both the primordial heat and the heat from radioactive decay). This heat accumulated below the crust and drove the mantle temperature back up again, until pockets in the crust and mantle began to melt. This triggered the onset of mare volcanism.

The first pockets to "remelt" and drive magma to the surface were the uranium, potassium and aluminum-rich lenses that had con-

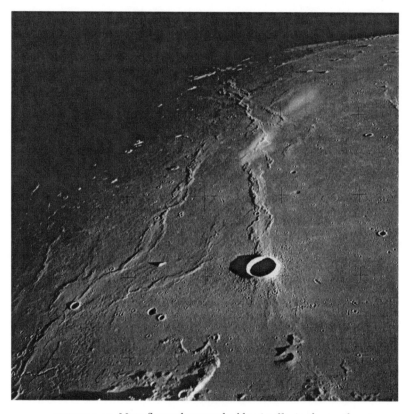

FIGURE 3.15 Mare flows photographed by *Apollo*, in the southeastern reaches of mare Imbrium. Upon cooling, the lava sheets contract and buckle to form wrinkle ridges, striking prominently across the image. Old impact craters are embayed by the lava (crescent at bottom). Credit: NASA, Goddard Space Flight Center (NSSDC).

gealed in the lower crust a few million years prior. Remobilized, they rose to the surface along the deep fractures that girdle impact basins. These are the KREEP basalts which the *Apollo 15* astronauts collected on the outskirts of the Imbrium basin, in the Apennine foothills, and were dated at 3.85 billion years.

More conventional basalts followed suit, coming from deeper down in the lunar mantle. They flooded most of the basins on the Moon's near side, starting about 3.8 billion years ago (*Apollo 11* samples, Sea of Tranquillity) and spanning hundreds of millions of years. The *Apollo 12* samples (Ocean of Storms) are the youngest lavas collected by the astronauts, coming in at 3.2 billion years.

Volcanic activity then sharply declined. The Moon had released most of its excess heat and the upper mantle generated smaller batches of magma that found it increasingly harder to reach the surface. There are a few lava flows in mare Imbrium that might be as young as 2 billion years. They were probably the last large eruptions to take place on the Moon. There have been hints of more recent events and even contemporary glows or *lunar transient phenomena* (namely over Alphonsus crater in 1958 and the Aristarchus plateau in 1963). Perhaps is there some gas that still reaches the surface and kicks up some dust (*Apollo* orbiters did detect small quantities of radon gas over the margins of some maria and over craters Aristarchus and Grimaldi). But whatever activity might still linger on, it is small and discrete enough to leave virtually no trace on the lunar surface.

In summary, the mare lava plains cover approximately 20% of the lunar surface. Each "sea" is comparable in size to a flood basalt province on Earth (mare Imbrium covers $500000\,km^2$, equal to the exposed area of the Deccan Traps of India). The total volume of lunar lava reaches close to ten million cubic kilometers.

However impressive, these figures represent a modest output, when averaged over billions of years of lunar history. During the peak period of mare volcanism, only one hundredth of a cubic kilometer of lava was extruded per year on average, which is comparable to the yearly output of a single volcano on Earth, like Vesuvius or Kilauea.

In comparison, the global output of all terrestrial volcanoes reaches three cubic kilometers per year – 300-fold the lunar figure. However, although the lunar average was low, the process still consisted of large eruptions, separated by thousands of years of quiet.

Single lava flows on the Moon are voluminous, and the large sinuous rilles indicate high discharge rates, perhaps as high as 100 000 cubic kilometers per year. In order to reconcile the instant and long term rates, one must assume that major eruptions on the Moon occurred on average only once every 100 000 years.

We can only imagine the majesty of these voluminous eruptions. In the lunar vacuum, very little gas is needed to make the magma froth and to drive lava fountains high above the vents. During their flight, the fine droplets of magma will chill into glass beads, like the green glass collected on the banks of Hadley Rille and the orange glass sampled at Taurus-Littrow. Because of the vacuum and low gravity on the Moon (1/6 g), these particles will fly large distances and fall back over a wide area, up to several kilometers around the vents. Spread out so thinly, volcanic ejecta fail to build any significant relief. Steep ash cones are not an option on the Moon.

LOOKING FOR VOLCANOES

On the Moon, many factors conspire to inhibit the growth of volcanic edifices, comparable to the broad shields, cones and domes that accompany igneous activity on Earth. The low gravity and near-vacuum help to disperse pyroclasts and bar them from building scoria cones and mounds around the erupting vents. Pyroclasts scatter instead to form flat blankets of glassy particles, known as *Dark Mantle deposits*.

Lava flows also spread far and wide, because of their low viscosity and high discharge rates, which lead to extensive lava plains, rather than to stacks of shorter overlapping flows. In addition, the absence of siliceous magma on the Moon rules out the possibility of thick pasty domes and lava plugs. Hopes to find such domes and ash flows on the Moon were dashed time and time again during the

Apollo missions, most notably when John Young and Charles Duke (*Apollo 16*) found that the light-colored plains and mounds around Descartes crater were impact debris rather than volcanic material.

One exception might be the Gruithuisen domes – a cluster of three steep hills in the northeastern reaches of the Ocean of Storms. We examine this unusual formation in the next chapter, as the target of a future field trip.

Despite the lack of spectacular mountains of lava, a number of small low-relief shields on the Moon do qualify as volcanoes. Several kilometers in diameter, a few hundred meters in height, these gentle

FIGURE 3.16 The Rumker Hills are a broad swell, capped by shallow shields, in the Ocean of Storms. This is one of the rare volcanic complexes visible on the Moon. Credit: NASA headquarters, Project *Apollo* Archive.

swells are visible under grazing lighting conditions. They are clustered along mare margins, as well as in the middle of the Ocean of Storms, which appears to have focused some of the most voluminous and long-lasting igneous activity on the Moon. Such clusters of volcanoes include the Rumker Hills and the Marius Hills, which boast 300 shallow shields and cones, together with meandering lava channels and lava tubes (we take a closer look at the Marius Hills in the next chapter).

Some features resemble cinder cones. Others resemble the small shields of Iceland and of the Snake River plain of Idaho. They were probably formed in a similar fashion, with lava rising through narrow conduits at a low discharge rate. Because of their modest volume, the lava flows cooled readily after exiting the vent and came to a stop after flowing only a short distance, piling up to build small shields.

Across the Moon, there could be as many as 500 to 600 of these shallow "pancakes" of lava and ash. They are a far cry from the towering features of the Earth, Mars and Venus, but as the only true volcanoes on the Moon, they will one day receive the visit of geologists from planet Earth.

4   A tour of lunar volcanoes

Hadley Rille

Latitude:	26° 5′ N
Longitude:	3° 40′ E
Length:	100 km
Width:	1200 m
Depth:	300–400 m
Estimated age:	3.7 billion years

FIGURE 4.1
Downstream section of Hadley Rille, curving around St. George impact crater. The *Apollo 15* landing site is in the bottom part of the frame. Credit: NASA.

FIGURE 4.2 Section of Hadley Rille visited by *Apollo 15*. See map opposite. Credit: NASA Headquarters, Project *Apollo* Archive.

Located on the edge of mare Imbrium (the Sea of Rains), Hadley Rille is an exciting destination for our first field trip to the Moon. It is a historical site, since it was first explored by the *Apollo 15* astronauts in 1971 (see pages 72–76).

The area contains two different units of lava: early aluminous basalt that crops out along the mountain front, and titanium basalt that later filled the basin. The rille itself is spectacular: snaking along the mountain front, it once carried great rivers of lava. There are also clusters of craters and mounds along the rille, which could be small volcanoes and are worth a close look.

Astronauts Dave Scott and Jim Irwin did a remarkable job studying the site and collecting 77 kg of rock samples during their three-day stay at Hadley Rille, in the summer of 1971. We can therefore visit the

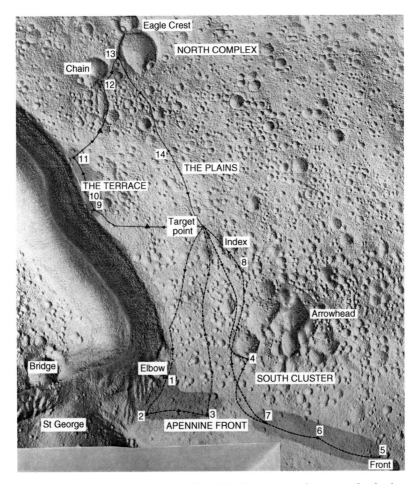

FIGURE 4.3 Map of the *Apollo 15* landing site, on the eastern bank of Hadley Rille, with suggested itineraries. The view from Elbow crater is shown in Fig. 4.4 and the wall of the rille at "The Terrace" is shown in Fig. 4.5. Credit: base map NASA/Defense Mapping Agency.

site with background knowledge, and pursue additional objectives that the astronauts were not able to accomplish for lack of time.

What might one day be known as the "*Apollo 15* Historical Park" is a stretch of mare plains wedged in between Hadley Rille to the west and the Apennine mountains to the east, in the southeastern corner of mare Imbrium.

The Apennine mountain range was uplifted by the violent asteroid collision that carved out the Imbrium basin 3.85 billion years ago – one of the last grand hits that rocked the near side of the Moon. The fractured crust channeled a first wave of magma to the surface, lining the newly-formed basin with a coating of aluminous basalt. This lower layer of lava is now visible only on the periphery of the basin, where it is known as the Apennine Bench Formation.

Hundreds of millions of years later, a new batch of magma made its way to the surface, probably through the same peripheral ring faults that girdle the basin. These mare basalts filled the Imbrium basin, reaching an aggregate thickness of 5000 to 6000 m, with central pockets up to 8000 m deep.

Most eruptions buried their eruptive vents and conduits. One exception is Hadley Rille. The channel originates in a crescent-shaped cleft in the foothills of the Apennine Mountains. It descends to the mare flats and snakes northward along the edge of the basin, hugging the mountain front. At the *Apollo 15* site, after curling around the foot of Hadley Delta (a 3500-m-tall mountain), it strikes northwest across the plains and reconnects with the mountain front at the foot of Mount Hadley (4500 m), resuming its winding course along the range front. Hadley Rille is over 100 km long, averages 1200 meters in width, and displays a V-shape cross-section with a rim-to-bottom depth of 300 to 400 m along most of its course, except where impact craters have collapsed the walls and filled in the channel. Based on the lava samples collected by the *Apollo* astronauts, Hadley Rille was active around 3.4 to 3.3 billion years ago.

Itinerary

We begin our itinerary at the *Apollo 15* landing site, roped off to preserve the historic footprints of Dave Scott and Jim Irwin. Prominently displayed are the abandoned lunar rover and science station, as well as the lunar module's descent stage wrapped in golden mylar – the platform from which the two astronauts blasted back into lunar orbit.

Like the two pioneers did on their first field day, back in 1971, we begin by riding our exploration rover to South George crater, in the foothills of Mount Hadley Delta. From this vantage point we enjoy a commanding view of the rille, striking northwest across the plains (see Fig. 4.4). Depending on the time of day (a lunar day lasts 28 Earth-days), shadows obscure parts of the channel. This is one of the best opportunities for photos, with good perspective and depth of field. It is easy to imagine the fiery river that swept and churned down the channel, a flow comparable to that of the Thames in London or the Hudson in New York City – on the order of one thousand cubic meters per second.

At our feet, where the channel takes a left-hand bend before heading northwest, we notice that the right bank stands about 30 m higher than the left bank. One interpretation is that the molten magma surged and overflowed to the right as it gushed into the left-turning bend, building a "plastering" of basalt on the outer bank.

This stop in the Apennine foothills is a good place for rock hunting. This is the Apennine Bench Formation, where we can collect good samples of the old aluminum-rich basalt (KREEP). We might also be fortunate enough to find some white sparkling chunks of anorthosite shed off the mountain front. These feldspar-rich rocks are remnants of the Moon's primeval crust and crystallized 4.5 billion years ago out of what was perhaps a global magma ocean covering the entire Moon. Lastly, our collection would not be complete without a spadeful of the emerald green glass that crops up here and there – droplets of volcanic spatter from the lava fountains that surged from nearby fissures.

The next leg of our field trip has us driving down the right bank of Hadley Rille, out to the terrace overlook of *Apollo 15*. From there, looking across the rille to the left bank, we can make out layering in the channel wall – a rare view of lava "bedrock" in place (see Fig. 4.5). At the very top, there is a unit of lava poking out of the rubble with joints dipping at a steep angle to the right. Underneath, another unit shows up in places, at least five meters thick, with distinct horizontal bedding. Below that, a thick talus blankets the rest of the slope,

FIGURE 4.4 Hadley Rille looking north, as seen from Elbow crater. The rille is 1500 m wide and 300 m deep. Credit: NASA Headquarters, Project *Apollo* Archive.

studded with boulders of all shapes and sizes. Although the slope is gentle – about 25° – we want to secure ourselves with a tether, before abseiling down to the outcrops and chipping away with our hammers. We will collect two varieties of basalt at Hadley Rille: a pyroxene basalt – the dominant type – and an olivine basalt found in the uppermost layer. Ages run from 3.3 to 3.4 billion years.

Our last stop will be at North Complex, to look at a cluster of hills and craters which were once believed to be volcanoes. In particular, there is a row of craters, called "the chain," that leads to the top

FIGURE 4.5 West wall of Hadley Rille, viewed from "The Terrace." Lava layers protrude through the talus rubble. Top unit shows joints dipping to the right. A thinner unit, at center of the image, shows horizontal bedding. Credit: NASA, courtesy of the NSSDC.

of the complex. Such crater chains can be rows of volcanic pits along a fissure, and we might find this indeed to be the case. But they can also be formed by grazing impacts, blasting ejecta down range into lines of "secondary" craters. It will be for us to solve the mystery . . .

Taurus-Littrow valley

Latitude:	20° 10′ N
Longitude:	30° 46′ E
Length:	50 km
Width:	7 km
Estimated age:	3.8–3.6 billion years

FIGURE 4.6 The valley of Taurus-Littrow, seen from the window of *Apollo 17*'s lunar module. South Massif in center frame, with its triangular avalanche deposit stretching out into the plains. Credit: NASA.

The *Apollo 17* landing site of Taurus-Littrow makes for a historic and instructive field-trip to the Moon. It offers us insight into ancient lunar history, as well as into mare volcanism. In particular, the site is well known for its colorful beads of orange glass that were formed in lava fountains.

The valley of Taurus-Littrow is a lava-filled *graben* (a faulted-down block) on the southeastern margin of the Sea of Serenity. The graben is radial to the 700-km-wide basin and surrounded by mountain ranges uplifted by the giant impact, around 3.9 billion years ago. The dark valley fill is part of a broader annulus of low-albedo material that rings the southeastern section of the Serenity basin. Before the *Apollo* landings, this annulus was believed to represent ash falls from relatively young volcanoes. In the middle of the valley, dark-halo craters were spotted from orbit, which resembled cinder cones and explosive pits. They were found instead, by the *Apollo 17* astronauts, to be young impact craters that dug down into the basalt layers, and sprayed dark rubble and glass beads onto the surface. The bulk of the lava pile in the valley, estimated to be 1.2 km thick, is a high-titanium basalt that was emplaced 3.7 billion years ago. Outside the Taurus-Littrow valley, towards the center of the sea of Serenity, there are younger basalts, less rich in titanium, that are 3.5 to 2.5 billion years old, according to crater counts. This trend from high-titanium to low-titanium can be explained if the source of the magma got deeper with time, migrating from the titanium-rich superficial layers to a lower mantle containing less titanium. This trend is not unexpected since the Moon experienced a progressive downward solidification of its upper mantle, and thus a migration of the partially molten zone to ever-increasing depths. The last eruptions at Taurus-Littrow – responsible for the orange glass – tapped a volatile-rich level in the mantle, nearly 500 km below the surface.

Itinerary

Our tour of Taurus-Littrow valley begins at the *Apollo 17* landing site, where the lunar module lower stage and science station are roped off

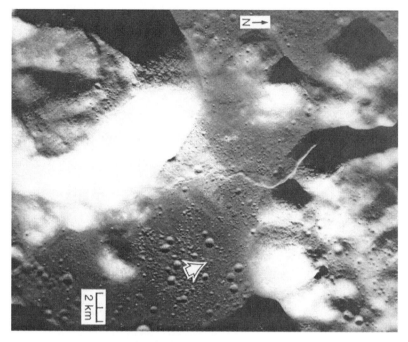

FIGURE 4.7 Overhead view of Taurus-Littrow valley. Notice scarp
running at top, ending beneath avalanche deposit. Matching map is on
facing page. Credit: NASA Headquarters, *Apollo* Lunar Surface Journal.

as a historic exhibit. After touring the site, charged with history, we
follow a route parallel to the original tracks of the astronauts, but far
enough so as not to spray dust over their historic footsteps.

Our first stop is at the foot of South Massif. It is part of the
Taurus mountain range that was uplifted by the Serenity impact,
approximately 3.9 billion years ago. The steep crustal block towers
2300 m above us and shows distinct layering, with a blue-gray unit
overlying a tan-gray unit towards the top. An avalanche scar sweeps
down the slope and extends a tongue of bright material, ten kilometers wide and twelve kilometers long, across the plains. Boulders
several meters across litter the sampling area, which tumbled down
from the upper stratified walls. Some boulders have a light-gray
friable matrix, easy to sample with a hammer. Others are harder and
frothy, with a blue-gray matrix full of vesicles. This was once partially

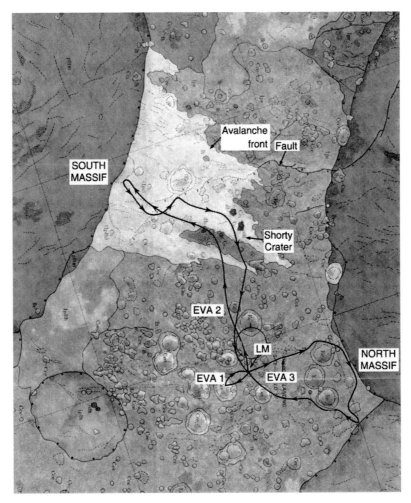

FIGURE 4.8 Map of the *Apollo 17* landing site in Taurus-Littrow valley, with itineraries followed by the astronauts. Credit: NASA/USGS, base map by D. H. Scott, B. K. Lucchitta and M. H. Carr.

molten and encloses a variety of rock fragments. Impact specialists will have a field day here and volcanologists will also land a few gems.

Some of the rock fragments in the breccia indeed date back to the earliest days of lunar history. They are light-colored, sparkling fragments of anorthosite and *norite* – the feldspar-rich "scum" of the primitive magma ocean that once covered the Moon (see pages

90–91). Other specimens are light green: olivine-rich *dunites*, full of iron and magnesium, which form at the bottom of magma chambers. Some of these samples might be as old as 4.5 billion years. There might also be chips of lava in the boulders, dating back to early mare volcanism. These old basalts are aluminum-rich.

To sample the mare basalts that later flooded the basins, we must descend to the dark floor of the valley, outside the avalanche blanket. On our way down the slope, we might stop to sample an extra boulder or two, or scoop up some light-colored avalanche soil. Cosmic-ray exposure shows the avalanche to be 109 million years old. It appears that the avalanche was caused by grazing blocks thrown out of the Tycho impact crater (80 km in diameter), 2000 km to the southwest. The age of the avalanche then reflects the age of Tycho crater.

Our next stop is Shorty crater, a young impact structure (19 million years old) that punched through the avalanche layer, the underlying soil, and down to the mare lava flows. Rocks are perched on the rim and walls of the 110-m-wide cavity, with benches of slumped material down to the hummocky floor.

On the southern rim of the crater, a large boulder marks the sampling spot of the *Apollo 17* astronauts, roped off to the public. The trench, which they dug in 1972, exhibits bright orange soil, believed at the time to be the fumarole deposit of a young volcano. It is made up of small beads of glass, rich in titanium. But although the glass is volcanic, Shorty crater is an impact crater. The orange glass is much older than the impact and dates back to the eruptions that filled the valley 3.6 billion years ago. The beads of glass were formed in lava fountains, and their layers were then buried by lava flows. Only recently did the Shorty impact blast the protective lid and shoot some of the beads up to the surface.

We limit our exploring to the northern half of Shorty crater, the side untouched by the *Apollo* astronauts. There is a streak of orange soil on the interior wall, a short distance from the rim, and we can reach it through careful sidestepping (a tether is recommended for safety). We want to pay attention to the boundaries of the orange

FIGURE 4.9 The bottom and northern wall of Shorty crater, where orange volcanic glass can be found. Credit: NASA Headquarters, Project *Apollo* Archive.

pocket and determine if it is emplaced along a fracture. Is the contrast sharp with the gray surrounding soil? Is there a color gradation – say yellow to crimson – from the edge to the heart of the deposit?

Driving a core tube through the glass layer should also help us to determine its vertical profile. We should be able to collect some underlying purple beads as well, which are iron-rich.

Moving farther down the slope, we look for chunks of basalt, ripped out of the underlying lava flows by the impact. We will probably find most samples to be coarse-grained basalt – the large crystals pointing to slow cooling in the midst of a rather thick flow. But there are also fine-grained varieties, belonging to thinner flows. We know from geophysics that the lava pile in the valley totals 1200 m in thickness, but the Shorty impact only sampled the top 100 m of the sequence.

A last stop on our tour of Taurus-Littrow is the north-trending fault line that crosses the valley floor, three kilometers west of Shorty crater. Named the Lincoln-Lee scarp, it resembles the wrinkle ridges that run across mare plains on orbital imagery. Up close, the scarp is steep, rising 80 m to the western section of the valley floor. Perhaps the eastern section on which we are standing sank down by that amount. Or else the western half was thrust over our eastern half, the scarp marking the front of the slab. Both kinds of motion are what you would expect from a cooling, contracting lava pile. Seismic profiling of the underground with small explosives and an array of seismometers would help clarify the issue.

Our field trip complete, as we drive back to the landing site, we get to enjoy a majestic view of the mountains towering over the valley floor and of a bright blue Earth hanging still over South Massif.

The Marius Hills

Latitude:	9° N–17° N
Longitude:	303° W–312° W
Area:	35 000 km^2
Estimated age:	3.3 billion years
Features:	262 domes
	59 cones
	20 sinuous rilles

FIGURE 4.10 The Marius Hills: a cluster of low shields amidst the wrinkle ridges of the Ocean of Storms. Image by *Lunar Orbiter 2*. Credit: NASA, Goddard Space Flight Center (NSSDC).

The Marius Hills consist of approximately 300 small cones and domes clustered on a plateau the size of Switzerland (35 000 km²). This is one of the few areas on the Moon that displays volcanic relief, albeit modest in height: most edifices are under 200 m tall. Sinuous rilles, interpreted to be lava channels and collapsed lava tubes, complete the picture.

The Marius Hills made it into the final cut of *Apollo* landing sites but lost out to the more spectacular Hadley Rille (*Apollo 15*) and Taurus-Littrow (*Apollo 17*). Had the last three *Apollo* missions not been cancelled, there is little doubt that Marius Hills would have

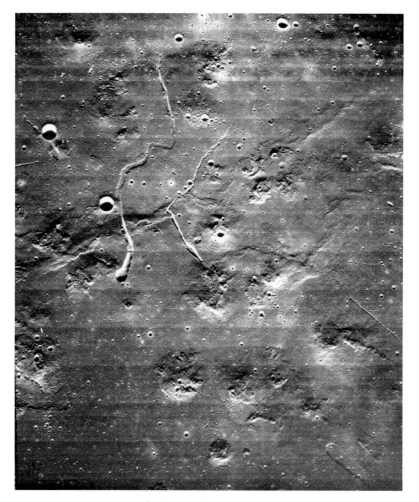

FIGURE 4.11 Overhead view of the Marius Hills: low shields, wrinkle ridges and two prominent lava channels. The landing spot of our excursion is the cluster of four shields, above and right of center. Image by *Lunar Orbiter 5*. Credit: NASA, Goddard Space Flight Center (NSSDC).

been visited in turn. In our field trip below, we will focus on an area in the Marius Hills that geologist Ronald Greeley proposed as an *Apollo* landing site.

The Marius Hills stand out as a broad plateau rising a couple hundred meters above the lava plains of the Ocean of Storms. They comprise 262 domes, up to 25 km in diameter and 500 m in height, which can best be described as lava shields. They also harbour 59 steeper cones, up to 3 km in diameter and 300 m in height, which probably were formed by gas-rich eruptions, spraying lava spatter around the vents. Finally, there are 20 sinuous rilles that emerge from craters or clefts and snake across the plateau and into the plains. These rilles are open lava channels and collapsed lava tubes.

The Marius Hills mark the source of abundant lava flows that spread through the Ocean of Storms. Known as the Hermann Formation, these flows appear to be low-titanium basalts (based on remote sensing from orbit) and crater counts pin down their age at roughly 3.3 billion years. The domes and cones probably grew late in the eruptive sequence, when activity waned and the lava piled up around the vents, instead of spreading far and wide across the plains. Hence, the Marius Hills offer us a rare opportunity to explore volcanic vents on the Moon.

Itinerary
In lunar orbit, we start our descent over the cratered highlands of the far side. As the near side comes into view, so does the vast smooth expanse of the Ocean of Storms, with the Marius Hills rising as a broad swell on the horizon. Meandering channels run down from the plateau into the plains, obviously the feeding conduits of much of the lava fields unfurling beneath our spacecraft. As the plateau grows closer, we aim for the northwestern edge, where four hills, each approximately 3 km wide and 200 m tall, mark the corners of a large square. We aim for a smooth lava field nested in the center.

On our way down we pass between the two western hills. They look like cinder or "spatter" cones with large horseshoe craters breaching their flanks. We can make out the dotted lineament of a partially collapsed lava tube, connecting the two volcanoes.

Once on the ground, after donning our spacesuits and deploying our rover, our first target is the southernmost volcanic hill, a mere two kilometers away. The hill is elongated in a horseshoe pattern –

FIGURE 4.12 Close-up of four Marius hills. Upper left corner of square is the breached cone that is visited in our itinerary. Image by *Lunar Orbiter 5*. North is to the right. Credit: NASA, Goddard Space Flight Center (NSSDC).

3.4 km long and 1.6 km wide – with a central trough nested between two ridges. Apparently, the eruption stretched out along a fissure, and the lava spatter piled up on two sides, without closing the northwestern and southeastern ends. As we have learned from similar breached craters on Earth, such open ends are caused by lava flows that carry away the falling scoria on their back, like conveyor belts, and prevent the full closure of the edifice. We progress on foot into the vast amphitheater, with the spatter walls rising 100 m on either side. The width of the valley – close to a kilometer at its mouth – reminds us of the enlarging effect caused by the low lunar gravity. Lava spatter, ejected by the escaping gases, fly farther away from the vent than on Earth, building wider and shallower flanks. We pause here to collect the frothy basalt spatter that should allow us to assess in the lab the quantity and nature of the propelling gases – most likely carbon monoxide.

After collecting our samples, we continue to the extremity of the narrowing amphitheater, progressing gently up slope until we reach the low wall that closes the southern end. Climbing to the top, we get a good view of the surrounding plateau, with gentle swells in the distance that look like buried fissures or wrinkle ridges caused by the buckling of the cooling lava. At the very foot of our cone, a rille is seen to emerge from the rubble and can be traced southward over a distance of approximately two kilometers. Rather than a large open channel, like Hadley Rille, it looks more like a collapsed lava tube, similar to the one that we flew over during landing. Its course is dis-

FIGURE 4.13 Will astronauts one day explore a lava tube on the Moon?
The Marius Hills might be the spot. Credit: photo by the author.

continuous – a dotted line – with open collapsed segments, separated
by others that are still roofed over by a crust of lava.

Walking down the outer flank of our fissure cone, we enter the
lava channel, picking our way through the jumble of rocks on the
floor. Perhaps will we be lucky enough to encounter a roofed-over
section that is still hollow, although the odds are very slim, given the
constant meteorite pounding that has shaken the rocks loose over the
past three billion years. But if we found it, how could we resist enter-
ing such a cave, despite the threat of a hidden crevice or a rock fall?

Perhaps in the permanent obscurity and coldness of the lava
tunnel will we discover coatings of volatiles – water ice, sulfur or
some exotic salt – frozen inside cracks and fissures. Despite the harsh
vacuum that turns them to vapor, the deposits might occasionally be
replenished by clouds of gas blowing in from nearby impacts. We will
also search the tube for the best-preserved chunks of lava, undamaged
by micrometeorites, which might still contain microscopic bubbles
of gas from their eruption.

After climbing out of the channel with our booty, we walk back to our rover and drive back to base camp to recharge our packs and break for lunch. We are in the Marius Hills for a while. There are so many cones, domes and rilles that call for a visit!

The Gruithuisen domes

FIGURE 4.14 The Gruithuisen domes, on the edge of the Ocean of Storms. Gruithuisen Gamma is to the left and the irregular mound of Gruithuisen Delta to the right. Image by *Apollo 15*. Credit: NASA Headquarters, Project *Apollo* Archive.

	Gruithuisen Gamma	*Gruithuisen Delta*
Latitude:	36° 30′ N	36° 10′ N
Longitude:	40° 35′ W	39° 20′ W
Diameter:	20 km (24 km × 18 km)	33 km × 14 km
Elevation:	1200 m	1600 m
Slope:	15–30°	15–30°
Estimated age:	3.7–3.8 billion years	3.7–3.8 billion years

Besides a few low shields in the mare plains (i.e. the Marius Hills), volcanic constructs are extremely rare on the Moon. The Gruithuisen domes, on the edge of the Ocean of Storms, are one such exception. They display a steep profile and an unusual spectral signature, which both point to an exceptional – and yet unsampled – form of lunar volcanism.

The Gruithuisen domes are a cluster of three hills – Gruithuisen Gamma, Delta and North-West – that straddle the boundary between the highlands and the mare plains, in the northern reaches of the Ocean of Storms.

The giant Imbrium impact, to the northeast, is responsible for shattering the crust in the area. It is no coincidence that the Gruithuisen domes lie precisely on one of the major fault rings that

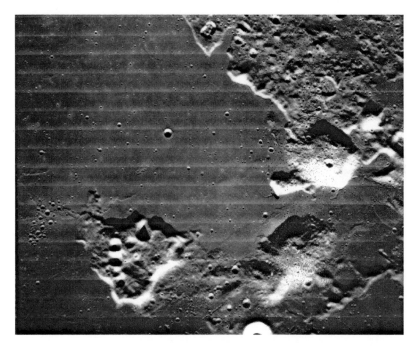

FIGURE 4.15 Overhead view of the Gruithuisen domes. Most circular dome (Gruithuisen Gamma) is topped by a pit: volcanic or impact crater? Note the two small sinuous rilles at foot of edifice. Image by *Lunar Orbiter 5*. Credit: NASA, Goddard Space Flight Center (NSSDC).

girdle the Imbrium basin. Such faults provided conduits for the magma to reach the surface.

The Gruithuisen Gamma dome is roughly circular in plan form, averages 20 km in diameter, reaches an elevation of 1200 m, and has steep slopes, a flat summit and a central pit. It is somewhat comparable in shape to the siliceous volcanoes of the Caribbean, like Guadeloupe's La Soufrière. Gruithuisen Gamma is flanked to the northwest by a smaller dome, 8 km in diameter. Southeast of the pair, separated by an encroachment of mare basalt, stands Gruithuisen Delta. More irregular in plan form (33 km × 14 km), it is the tallest volcano of the group and peaks 1600 m above the plains.

All domes display a reddish spectral signature that makes them stand out against the background. Such red spots on the Moon are believed to represent exposures of primitive, potassium-rich lava (KREEP) that predates the mainstream mare basalt. Remote-sensing by the *Clementine* and *Lunar Prospector* probes indicates a low iron content (<10%) and very low titanium (<1%). Despite the steep, viscous aspect of the volcanic constructs, there is no evidence that the lava is silica-rich. It need not be dacite or andesite to build a steep mound. Basalt, if erupted at a low rate, cools quickly and behaves viscously as well.

The Gruithuisen domes were formed shortly after the major impact that excavated the Imbrium basin, and their age is bracketed between 3.85 and 3.7 billion years ago. Mare lava flows later flooded the basins and surrounded the domes.

Itinerary

For a field trip to Gruithuisen domes, the safest approach is from the south, flying over the Ocean of Storms and coming in for a landing on the mare plains at the foot of Gruithuisen Gamma.

From our landing spot, the road to the top of Gruithuisen Gamma is a difficult hike up a steep slope that occasionally reaches 30 degrees – comparable to the steepest volcano slopes on Earth. In

FIGURE 4.16 Close-up of Gruithuisen Gamma, the target of our
excursion. Cone-shaped "North-West" hill is in the background. Credit:
NASA Headquarters, Project *Apollo* Archive.

our bulky spacesuits, despite the low gravity, we feel the strain of the
climb. There is no major obstacle but the volcano flank is hummocky
– the result of overlapping lava flows and landslide deposits.

As we pick our way up the slope, we are reminded how difficult
it is to do geology on the Moon, or on anything older than 3.7 or 3.8
billion years for that matter, when impacts were so numerous that
they shattered and mangled lava flows and other geological units. We
can scoop up some of the pulverized *regolith* – the lunar "soil" – that
gives off the distinctive reddish hue of the Gruithuisen area. Perhaps
the red color is due to the larger proportion of volcanic glass, spread
across the domes by gas-rich eruptions of ash.

Young impact craters might have tossed unaltered pieces of
rock from deep under onto the surface, and we will be on the lookout
for such pristine samples of lava. Will they turn out to be primitive,

aluminous basalt (KREEP) as expected, or something more evolved and siliceous like andesite or dacite?

After a seven-kilometer hike up the steep talus we reach the smooth, flat summit of the dome, 1200 m above the mare plains. In the distance, on the rolling horizon, we can make out the bumpy profile of our sister volcano Gruithuisen Delta, 30 km to the east. After catching our breath, we walk towards the location of the summit pit. Although the summit is rather flat, our progress is still hampered by the numerous impact craters that churn up the landscape. We look out for thick pockets of volcanic fall-out – layers of glass beads – which would attest to gas-rich pyroclastic eruptions at the summit.

The summit crater is a mystery in itself. More than two kilometers across – the size of Kilauea crater in Hawaii – could it be the true pit of the volcano, surviving three and a half billion years of erosion? Or is it yet another impact crater, located by chance at the center of the dome? It is for us to tell.

The Aristarchus plateau and Schroeter's valley

The Aristarchus plateau

Latitude:	21°N to 25°N
Longitude:	309°W to 313°W
Area:	40000 km²
Features:	36 sinuous rilles
Estimated age:	3.6 billion years

Schroeter's valley

Length:	160 km
Width (max):	11 km
Depth (max):	1000 m

FIGURE 4.17 The Aristarchus plateau viewed from *Apollo 15*. The Schroeter sinuous rille emerges from Herodotus crater (feature known as "The Cobra Head"). Credit: NASA.

The Aristarchus plateau is an island of hummocky terrain, surrounded by younger basalts in the middle of the Ocean of Storms. It boasts two sizeable impact craters: the older Herodotus ring, flooded with lava, and the much younger Aristarchus crater that spreads a hummocky blanket across the plateau. It also harbours a dense population of volcanic rilles (36), including the widest volcanic rille on the Moon: Schroeter's valley.

The plateau's volcanic nature was discussed at the dawn of the space age when Russian astronomer Kozyrev reported a luminescent phenomenon over the area in 1961, and astronomers at Lowell

Observatory spotted a red glow over Aristarchus crater in 1967. Instruments aboard the *Apollo* orbiters detected "whiffs" of radon gas – the decay product of uranium – as they flew over the site in the early 1970s. Although it is established that there is no recent volcanism at Aristarchus, a large impact crater does fault the basement of an extinct volcanic province. It is possible that gas from deep down in the mantle still manages to reach the surface in the area.

The Aristarchus plateau is an ancient crustal block, rectangular in shape and over 200 km across, which was uplifted and fractured by the giant Imbrian impact, 3.85 billion years ago. The tilted block rises 2 km above the Ocean of Storms at its southeastern margin and slopes down to basin level in the northwest. The fracturing that accompanied the block's uplift created pathways that the magma would later follow to the surface.

The major phase of volcanism on the Aristarchus plateau took place around 3.6 billion years ago and involved the eruption of vast quantities of lava known as the Telemann Formation. These low-titanium basalts (as indicated by their spectral signature) flowed into the surrounding basin and account for a large fraction of the mare plains in the northern Ocean of Storms. Many flows were emplaced by sinuous rilles that channeled the magma from the plateau down to the plains. There are 36 such rilles, including Schroeter's valley (the site of our field trip). Many have the plan form of tadpoles, with large oval sources and narrow tails. The Aristarchus plateau and some of the surrounding landscape are covered by dark mantling material, reddish brown in color, which is interpreted to be iron-rich glass, sprayed by lava fountains at the onset of major eruptions.

Much later in the history of the plateau, a major impact excavated Aristarchus crater in its southeastern corner. The bright terraced arena is 42 km in diameter and its age is estimated to be 450 million years. The impact straddles the plateau and the mare plains, and sprayed bluish ejecta from the crustal roots of the plateau to the northwest, and ejecta of reddish mare basalt to the southeast.

FIGURE 4.18 Schroeter's valley lava channel. Flow is from left to right.
Credit: NASA Headquarters, Project *Apollo* Archive.

Despite the large impact, the older volcanic features are well
preserved, making the Aristarchus plateau a choice location for our
last field trip to the Moon. In fact, it was scheduled for *Apollo 18*,
before the last three missions were cancelled. Our main objective is
the sinuous rille that snakes down the northern slope of the plateau:
Schroeter's valley.

Itinerary

From orbit, the Aristarchus plateau is a dazzling sight, rising above
the Ocean of Storms. Its two impact craters lie side by side: the old

FIGURE 4.19 The upstream section of Schroeter's valley. The lava rille emerges from an alcove shadowed by tall cliffs. Credit: NASA Headquarters, from *Apollo over the Moon: A View from Orbit* (NASA SP-362), eds. H. Masursky, C. W. Colton and F. El-Baz.

Herodotus ring, flooded by lava, and the much younger Aristarchus, boasting bright streaks of ejecta.

We fly in from the north, over the lava plains that surround the plateau. They merge into a hummocky slope, riddled with impact craters, including many "secondaries" from Aristarchus. One foothill at the base of the plateau, smooth and lens-shaped, might be a rare example of a lava shield. Otherwise, the main expression of volcanism on the plateau is the array of tadpole-looking rilles. As we fly over the hummocky rise of Aristarchus crater, the rilles grow more numerous, flowing down to the east. Instead, we take a turn to the west, aiming for the saddle that separates Aristarchus and Herodotus craters. Halfway up the slope, a huge cleft bites into the plateau, unfurling into a 160-km-long rille that snakes away to the northwest: Schroeter's valley.

The wide source area of the valley is known as the "Cobra Head" for reasons obvious. The vast amphitheatre, over 10 km wide, reminds us of sector collapses that carve into many volcanoes on Earth, like the Valle del Bove that cuts through the flank of Mount Etna. We bring our lander over the edge of the cleft and plunge 1000 meters down to its floor. Exceptionally, this is a noon landing – the sun is high in the sky – so that the target area is not hidden in shadow. Within the vast Cobra Head, there is a rille within the rille – a nested, meandering channel. Our lander settles in a cloud of dust, at the source of the channel.

As we step out into the open, we take in the breathtaking view. The walls towering above us are scarred by avalanche tracks. Debris shed from the cliffs stretch across the valley floor and spill over into the central rille. We head for this inner rille, which conveyed the last flows of lava down Schroeter's valley, close to 3.6 billion years ago, and attempt to track back to the source vent, amidst the rubble. Above our heads, the layering in the head wall appears to shine blue – certainly layers of highland crust that are rich in feldspar. We sample detached blocks of crust that fell from above, mixed with slabs of basalt that rose from below, out of the buried vent. We also set up a particle detector, to record any evidence of degassing from the deep lunar crust, through the choked lava pathway.

Patches of reddish soil are widespread in the valley. We scoop spadefuls of the iron-rich glass to bring back to the lab – most certainly beads of magma that rained out of lava fountains on the scene of the eruption. The Cobra Head will be our playground for the next

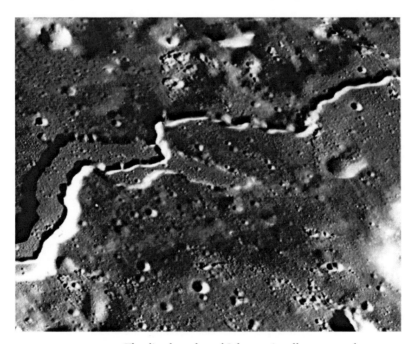

FIGURE 4.20 The distal reaches of Schroeter's valley, target of an extended field trip. Credit: NASA Headquarters, from *Apollo over the Moon: A View from Orbit* (NASA SP-362), eds. H. Masursky, C. W. Colton and F. El-Baz.

few days. We can elect to stay in the source area, or else pack our gear and drive down the valley, 160 km to its narrowing outlet in the Ocean of Storms. But we only have four or five days of good light left, before the Sun disappears behind the cliffs and plunges Schroeter's valley into the long lunar night.

5 Volcanism on Mars

Mars is a spectacular showcase of planetary volcanism. It boasts the two largest volcanoes in the Solar System – Olympus Mons and Alba Patera – and a host of other shields, calderas and lava fields. These giant volcanoes were mostly unsuspected before the space age, when telescopes pointed at Mars showed little more than a shimmering red disk with bright polar caps and mysterious dark markings. It took the close scrutiny of space probes to unravel the marvels of the red planet. Orbiters beamed back stunning views of volcanoes, canyons and channels carved by running water. Landers touched down on the smooth lava plains and analyzed the rocks and soil. Moreover, although no space probe is expected to bring back Martian samples to Earth before 2014 at the earliest, several dozen rocks from Mars have already landed in our labs – chunks of lava that were kicked off the planet by large impacts and fell to Earth as Martian meteorites. They bring an interesting contribution to the study of Martian volcanism.

At the onset of the third millenium, a wealth of new data is pouring in, returned by the latest generation of space probes, namely *Mars Global Surveyor* (1997), *Mars Odyssey* (2001), *Mars Express* and the *Mars Exploration Rovers* (2004). By bringing Martian volcanism into sharper focus, they are challenging many of our assumptions and turning theories on end. Volcanic deposits – lava and ash – appear to be more extensive than previously thought, mantling the plains and jutting out of canyon walls. Small cones, pits and domes pepper the landscape on close-up imagery. Even the giant volcanoes are being reappraised in terms of shape, elevation and eruptive behavior, and most surprisingly they are now looking a lot younger.

It was once believed that Martian volcanoes had shut down *billions* of years ago. Close-up imagery taken by *Mars Global Surveyor* now show many of the lava fields to be remarkably free of impact

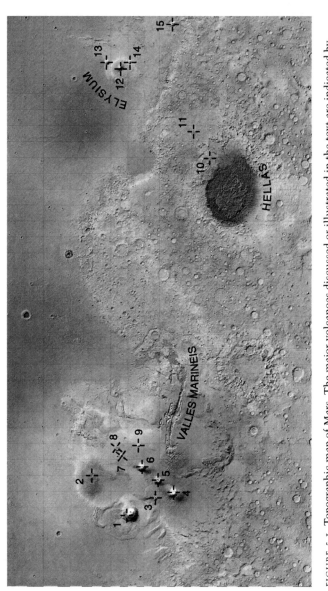

FIGURE 5.1 Topographic map of Mars. The major volcanoes discussed or illustrated in the text are indicated by numbers. (1) Olympus Mons; (2) Alba Patera; (3) Biblis Patera and Ulysses Patera; (4) Arsia Mons; (5) Pavonis Mons; (6) Ascraeus Mons; (7) Uranius Patera; (8) Ceraunius Tholus and Uranius Tholus; (9) Tharsis Tholus; (10) Hadriaca Patera; (11) Tyrrhena Patera; (12) Elysium Mons; (13) Hecates Tholus; (14) Albor Tholus; (15) Apollinaris Patera. Credit: NASA/MGS/MOLA.

craters, suggesting much younger ages. Several lava flows on Olympus Mons and in the Elysium basin appear to be little more than a few *million* years old, a one thousand-fold reduction in age. They imply that volcanism is still an ongoing process on Mars and that more eruptions might take place in the future.

Last but not least, the importance and longevity of Martian volcanism has major implications for the emergence of life on the red planet. Voluminous eruptions in the past released substantial amounts of gas into the atmosphere, including water vapor. Oceans and lakes probably filled the basins, craters and canyons. Hydrothermal circulation around hot vents and fumaroles might well have created favorable conditions for the emergence of life, as it presumably did on the early Earth. Indeed, one of the Martian meteorites – ALH 84001 – contains organic matter and putative microfossils inside fractures where fluids once circulated, although the findings are highly controversial and can be explained away by non-biological processes or from contamination by terrestrial rather than Martian bacteria.

Be it as it may, volcanoes on Mars will be targeted by robotic probes and manned expeditions in the future, not only because they will provide insight into the geological history of Mars, but also because they could shed light on the mystery of life itself. There might even be mineral ores in the fissures and dikes of Martian volcanoes, which future colonies will want to exploit, as well as geothermal fields that will supply heat and water to their settlements.

For all these reasons, astronauts are bound to hike one day the volcanoes of Mars. It is only a question of time.

THE DISCOVERY OF MARS

Mars is the fourth planet away from the Sun, circling just outside the Earth on an elliptical orbit that it covers in 687 days – the length of a Martian year. On our inner, faster orbit, we overtake Mars every 26 months, when we come as close as 55 million kilometers to the red planet, as we did in 2003. Mars then shines as a bright red star in our night sky, but shows few details in even the most powerful telescopes.

Mars is a small planet. Its diameter is 6794 km at the equator, and its area amounts to a quarter of the Earth's surface – which compares to the area of all our continents put together. By virtue of its low mass and volume, Mars was long thought to be a dead planet, much like the Moon. The consensus was that its internal heat had bled away at a fast pace, shutting down volcanism early in its history. That feeling was reinforced when the first spacecraft to fly by the red planet – *Mariner 4* in 1965 and *Mariner 6* and *7* in 1969 – beamed back fuzzy images of desolate landscapes, riddled with impact craters. The atmosphere turned out to be desperately thin – ten hectopascals (millibars) of pressure at best – and to be made up mainly of carbon dioxide, with only a trace of water vapor. Mars was a cold, dry and vacuum-like desert.

The first *Mariners* had only scanned three narrow swaths of imagery across the red planet. For a global coverage of Mars, geologists had to wait for *Mariner 9*, the first probe to achieve Martian orbit on November 14, 1971. Its mapping mission got off to a bad start. A global dust storm enveloped the planet in a thick opaque shroud, and the early images returned by *Mariner 9* were completely featureless. Flight controllers at Pasadena's Jet Propulsion Laboratory decided to shut down the cameras and wait out the storm – a wait that grew into riveting suspense for the scientists.

As the winds abated and the dust began to settle, four dark spots emerged from the cloud tops. They progressively came into focus as large craters, tens of kilometers in diameter. They could not be ordinary impact craters, like those spotted by previous probes. Since they were the first features to poke through the settling dust, they had to be calderas sitting atop giant volcanoes.

Calderas they were. As the dust veil dropped to the ground, the four volcanoes appeared in all their glory, straddling the equator between 100°W and 140°W. Three features, lined up on a NE–SW axis, were christened Ascraeus Mons, Pavonis Mons and Arsia Mons (from north to south). The fourth feature, detached to the northwest, matched a bright spot that astronomers had long noted on the

FIGURE 5.2 Olympus Mons is the tallest volcano in the Solar System, peaking 22000 m above the plains. The shield is over 500 km wide, with a steep cliff rimming the base. The summit caldera is 80 km wide. North is to the upper right. Credit: *Viking* image, NASA/JPL, courtesy of Brown University.

Martian disk. They had named the white spot Nix Olympica, "the snows of Olympus." Apparently, the volcano created cloud patterns that were visible from Earth, and the edifice aptly inherited the name Olympus Mons.

GIANT VOLCANOES

The volcanoes of Mars were a startling surprise to geologists. After failing to find sizeable edifices on the Moon, they were met with over-sized shields on the red planet. Tentative altimetry readings by *Mariner 9* hinted that the summits peaked 20000 meters above "sea level" on Mars – and even 27000m in the case of Olympus Mons. More accurate measurements, taken in the late 1990s by *Mars Global Surveyor*, would shave a few kilometers off these early estimates.

Olympus, Ascraeus, Pavonis and Arsia *Montes* (the plural of *Mons* in Latin) are extremely large. The shield of Olympus Mons averages 600km in diameter and covers an area the size of Arizona. By comparison, the largest volcano on Earth – Mauna Loa in the Hawaiian Islands – does not exceed 150km in diameter (as measured on the seafloor) and covers only a tenth of the area claimed by Olympus Mons.

Everything about the Tharsis foursome is off the scale with respect to their Earthly counterparts. The summit calderas are close to 100km in diameter and are bound by steep cliffs that drop thousands of meters down to their lava-covered floors. Lava flows, which on Earth rarely exceed 100km in length, reach run-out distances of 1000km and more on the Tharsis plateau.

The spectacular scale of Martian volcanoes was not the only mystery that puzzled scientists. There was also the question of age. How recent were these volcanoes? Were they still active? Could they still harbor eruptions? Or were they relict landforms, frozen in time, that had been extinct for billions of years?

Mars was a curious place. Much of the planet looked like the Moon. Saturated with impact craters, this terrain could not be much younger than four billion years. The Tharsis plateau and the northern plains, on the other hand, were much less disturbed by impacts. But assigning ages to these areas, based on crater counts, turned out to be a rather tricky exercise. Ages on the Moon were well constrained because astronauts had recovered rock samples at various landing sites, which were then dated in the laboratory. Therefore, scientists

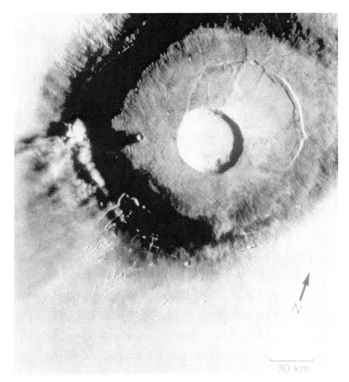

FIGURE 5.3 Pavonis Mons is the central volcano of the Tharsis plateau (flanked by Arsia Mons to the southwest and Ascraeus Mons to the northeast). It rises 14 000 m above the reference level. Its 90-km-wide central caldera is rimmed by a wider, more ancient collapse ring to the northeast. Note the clouds hugging the flank of the volcano to the southwest. Credit: *Viking* image, NASA/JPL, in *Volcanoes on Mars* slide set, compiled by James Zimbelman, courtesy of the Lunar and Planetary Institute.

could equate a terrain that bore a certain number of impact craters with an absolute age. They possessed a reliable scale.

On Mars, no one knew if the cratering rate had been the same as on the Moon and no "ground truth" – i.e. datable rocks – was available to calibrate the scale. Depending on their initial assumptions and crater-counting methods, scientists came up with very different estimates for the age of Martian volcanoes. One team found the giant shields of Tharsis to be three billion years old, while another deduced much younger ages for the least cratered areas: 700 million years for

Arsia Mons, and a mere 100 million years for Olympus Mons. As we shall see, independent evidence from Martian meteorites and more detailed imagery from later probes would support the younger ages.

THARSIS AND ALBA PATERA

While the giant shields of Tharsis were controversial in age, there was little doubt that smaller volcanoes on the plateau were significantly older. These are characterized by a large caldera perched on a shield of limited proportions. Arguably, they are but the summit sections of what were once much larger shields, now surrounded and buried by more recent lava flows. Today, we are only seeing the tips of these "lavabergs."

Old volcanoes on Mars are classified as *paterae* ("saucers" in Latin) if they appear rather flat, or *tholi* ("hills" in Latin) if they have steeper flanks. They tend to occur in clusters. The pair Biblis and Ulysses Paterae poke their calderas through the lava fields west of Pavonis Mons. The triplet Ceraunius Tholus, Uranius Tholus and Uranius Patera are located northeast of Ascraeus Mons, in line with the giant volcanoes, as if they were part of the same regional rift zone. Solitary shields are also present, notably Jovis Tholus and Tharsis Tholus, respectively on the northwestern and eastern margins of the Tharsis plateau. Each one of these volcanoes has a distinct shape and personality.

The Tharsis plateau hosts most of the volcanic activity on Mars, boasting four giant shields and seven lesser constructs. It is flanked to the northwest by an unusual feature that can be treated as an entire volcanic province onto itself: the gigantic shield of Alba Patera.

While Olympus Mons is the tallest volcano in the Solar System, Alba Patera is the largest, but it commands much less attention because its slopes are very gentle and its summit caldera shallow and discrete. Alba's shallow shield of lava is estimated to be at least 1000 kilometers in diameter – as large as the State of Texas – and its influence, in the form of ash falls and lava flows, extends perhaps to twice that distance. But it rises only 6000 m, so that its slopes average less than one degree. For each kilometer hiked towards the summit, an astronaut would rise less than ten meters.

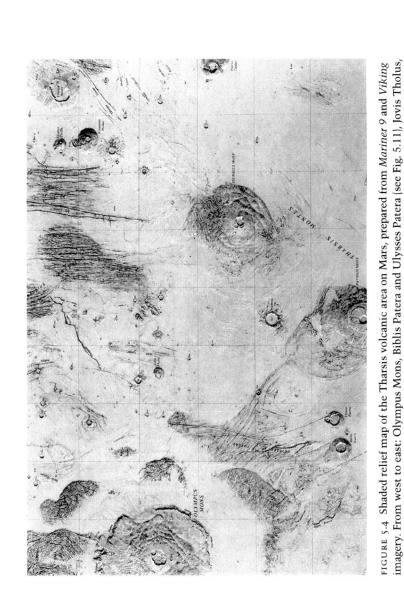

FIGURE 5.4 Shaded relief map of the Tharsis volcanic area on Mars, prepared from *Mariner 9* and *Viking* imagery. From west to east: Olympus Mons, Biblis Patera and Ulysses Patera (see Fig. 5.11), Jovis Tholus, Pavonis and Ascraeus Montes, Uranius Tholus (see Fig. 5.12) and Ceraunius Tholus (see Fig. 5.20), Uranius Patera and Tharsis Tholus (see Fig. 5.13). Credit: U. S. Geological Survey.

Alba Patera certainly took a considerable amount of time to grow to its present size. Its volume reaches two million cubic kilometers, the equivalent of a large igneous province on Earth, like the Siberian or the Deccan Trapps. Crater counts suggest that its growth took place midway through the planet's history, spanning the period 3 to 1.5 billion years ago, before activity shifted to the giant shields of Tharsis.

Alba's eruptive behavior also sets it apart. Some areas on the volcano lack lava flows and are furrowed instead by sinuous channels, like fields of ash sapped by flowing groundwater. If this is truly the case, then the middle epoch of Martian history (known as the *Hesperian* era) was marked by violent eruptions, fueled by water vapor that propelled clouds of ash into the atmosphere and sent pyroclastic flows rushing down the slopes.

ELYSIUM AND HELLAS

Evidence of a water-rich past and of explosive eruptions occurs in two other volcanic regions on Mars: the Elysium rise and the Hellas basin.

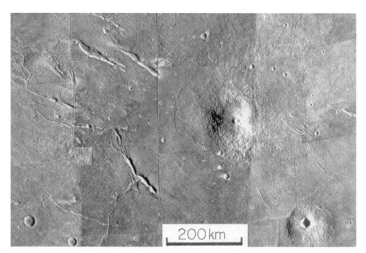

FIGURE 5.5 The Elysium rise comprises three major volcanoes. The southernmost two are Elysium Mons (top); and Albor Tholus (bottom, right of scale). The smooth upper flanks of Elysium Mons might be due to blankets of fine ash. Note the fault system (graben) trending northwest (left half of image). Credit: *Viking* photomosaic, NASA/JPL, courtesy of Brown University.

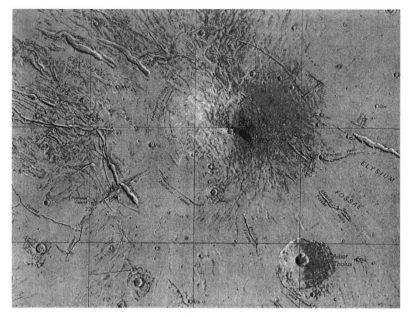

FIGURE 5.6 Shaded relief map of the southern part of the Elysium rise, prepared from *Mariner 9* and *Viking* imagery. Compare with Fig. 5.5 on facing page. Note the concentric graben around Elysium Mons (Stygis and Zephyrus Fossae), caused by the flexing of the crust under the volcanic load. Credit: U. S. Geological Survey.

Smaller than Tharsis, lying 2000 km to the west, Elysium is Mars' second largest volcanic province. It boasts three major shields, comparable to Earth's largest volcanoes. Albor Tholus lies in the south, a stubby low shield with a wide caldera. Elysium Mons stands in the middle, by far the largest of the trio, towering 14 km above the plains. Hecates Tholus rises in the north and boasts radial channels and a smooth ash field near the summit – evidence of both water run-off and explosive eruptions. Because they lack large impact craters, the Elysium shields appear to be "middle-aged," averaging perhaps one to two billion years. But south of the Elysium rise, there are vast lava fields with open fissures, low shields and broad unspoiled flows, which rank among the youngest units on Mars. Like several areas on Tharsis, they might still harbor eruptions to this day.

In contrast, activity long shut down in the third volcanic

province of Mars: the fractured rim of the Hellas impact basin in the southern hemisphere. Deep-seated faults funnelled magma to the surface in the distant past, when volatiles like water were abundant in the Martian environment. Hadriaca and Tyrrhena Paterae, perched on the eastern rim, and Amphitrites Patera on the southern border each consists of a shallow shield, topped by a large caldera and gouged by wide channels that radiate from the summit like the spokes of a giant bicycle wheel. Lava flows are conspicuously lacking on these volcanoes, the consensus being that their deeply eroded flanks are made of friable ash rather than hard bedrock. Abundant water, soaked up by the ash, allegedly carved the channels through surface run-off or subterraneous flow.

ERUPTIONS ON MARS

From our brief roundup of Martian volcanoes, we can gain much insight into the original conditions that govern igneous activity on the red planet.

The first striking characteristic of Martian volcanoes is their great size and longevity. The giant shields of Tharsis and to a lesser extent the volcanoes of Elysium and Hellas pack in many times more lava than the largest volcanoes on Earth, and yet Mars is a smaller planet. The reason lies in the planets' internal make-up. Because it is larger and hotter, the Earth has a nearly molten mantle, overlain by a very thin crust – less than 10 km thick under the ocean basins. The hot mantle churns over in great convection loops, and rips the overlying rigid crust into tectonic plates that shuffle around the globe. Where deep-seated "hot spots" send plumes of molten material to the surface, the magma starts piling up on the crust to build a volcano, but because the crust is drifting, the volcano eventually breaks away from its "plumbing system" and goes extinct. In its place, a new volcano starts to grow, so that a hot spot under a mobile plate creates a string of medium-sized volcanoes – in assembly line fashion – rather than one giant shield.

On the smaller, faster-cooling red planet, the crust has solidified and thickened to the extent that it remains stationary with respect to the underlying hot spots. When a plume of magma erupts at the surface, it will build a single volcano in one place for as long as the supply lasts – typically hundreds of millions of years. Not surprisingly, then, the volume of one Martian shield like Olympus Mons or Alba Patera is comparable to that of an entire chain of volcanoes on Earth, such as the Hawaiian Emperor chain that stretches across the Pacific and spans nearly 100 million years of hot spot activity.

Another characteristic of Martian volcanoes is the large size of

FIGURE 5.7 Tyrrhena Patera is an ancient volcano on the rim of the Hellas basin. It is nicknamed "Dandelion" on account of its petal-like channels, fanning out from the central caldera. The channel network is attributed to water flow, sapping its porous ash layers. Credit: NASA, *Viking*/JPL (Dr. Michael H. Carr, *Viking* Experiment Team Leader), courtesy of the National Space Science Data Center, through the World Data Center-A for Rockets and Satellites.

their calderas and the long run-out distances of their lava flows. This is principally due to the lower gravity on Mars: 0.38 g or about one third of Earth's gravity. Gravity controls to some extent how solid rock behaves under stress. When the crust stretches and fails in a low gravity field, as it does on Mars, the fissures that open up are wider than on Earth and can funnel larger amounts of magma towards the surface. Magma chambers grow correspondingly larger and when they empty out and collapse, they yield larger calderas.

Eruption rates also tend to be higher on Mars, since greater quantities of lava can flow out of the wider fissures. These larger volumes guarantee a better retention of heat. Because it stays hot and molten for a longer period of time, lava on Mars travels greater distances than it does on Earth before cooling and slowing to a halt.

Explosive eruptions are also affected by the unique set of conditions that exist on Mars. The atmospheric pressure is so low – six hectopascals in the plains and less than three atop the tallest volcanoes – that any gas bubbles trapped in the magma will undergo tremendous expansion upon reaching the surface. As a result, it will take lesser amounts of dissolved volatiles in Martian (versus terrestrial) magma for the mixture to foam and shoot out explosively from the vent, rising as hawaiian lava fountains, plinian ash clouds or pelean base surges.

On Earth, the confining pressure of the atmosphere requires that a magma contains close to 1% (by weight) of volatiles for it to spray upwards as a lava fountain. About 3% to 4% are needed for the bubbles to grow large enough to blow the magma to shreds and create an ash cloud. On Mars, the thresholds for such disruptive behavior are much lower: respectively 0.03% and 0.2% volatiles by weight. Twenty to thirty times less gas is needed on Mars to achieve similar results.

The consequences are profound. On Earth, basaltic magma rarely contains enough volatiles to blow up into ash clouds. Only at the onset of an eruption is there enough gas accumulated at the top of the magma column to drive lava fountains. It takes a more silicic

FIGURE 5.8 The caldera of Olympus Mons contains a variety of floor levels and collapse pits that testify to a shifting magma plume below the volcano. Credit: NASA, *Viking*/JPL (Dr. Michael H. Carr, *Viking* Experiment Team Leader), courtesy of the National Space Science Data Center, through the World Data Center-A for Rockets and Satellites.

magma, which can store more volatiles, to achieve an ash explosion. Therefore, on our planet, plinian and pelean ash clouds are mostly restricted to subduction-type volcanoes with a silica-rich chemistry, such as andesite, dacite and rhyolite domes.

On Mars, however, the identification of ash layers around the Hellas Paterae, Elysium's Hecates Tholus or the Tharsis shields does not necessarily mean that their magma is siliceous. The modest quantities of volatiles typical of basalt are sufficient to blow up the erupting magma into pulverized ash. Hence, geologists cannot simply rely on the aspect of volcanic deposits to deduce the chemistry of Martian lavas.

Low pressure and gravity have other consequences. Greater expansion of bubbles in the molten froth will stretch and break up their boundary walls into thinner shards of glass. For a given amount of volatiles, particles will thus be finer on Mars than on Earth, mostly millimetric to micrometric in size.

Emerging into near vacuum, the gas will also accelerate to higher velocities. Exhaust speeds out of volcanic vents will be one-and-a-half times greater on Mars than on Earth and will often exceed 500 m/s. Such supersonic speeds do not mean that larger "bombs" of lava will be lofted out of erupting vents. On the contrary, the gas is so rarefied that despite its greater velocity, its lift force will remain modest and it will propel only fine ash into the Martian sky. However, these clouds will be lofted to great heights. Coupled with the low gravity of the red planet (0.38 g), high exhaust velocities mean that a cloud will rise almost twice as high on Mars than on Earth, reaching altitudes of 50 km. And because they rise higher, eruption clouds should also spread wider, showering ash over larger areas than on Earth and building shallow edifices rather than steep cones and domes.

But this lofting mechanism only works to a point. If eruption rates are too high and propel massive amounts of matter out of the vent, the density of the ash column will quickly supersede the buoyancy of the thin atmosphere. Instead of rising, the erupting plume will collapse as a gravity-driven pyroclastic flow on the flanks of the volcano. This "collapse threshold" is reached for mass eruption rates much lower on Mars than on Earth, so that pyroclastic flows ought to be fairly frequent on the red planet.

Hence, theoretical considerations suggest that Martian eruptions will be tremendously diverse and mix quiet lava flows, lava fountains, ash clouds and pyroclastic flows. What we need next, to better understand Martian volcanism, is information on the chemical and mineralogical make-up of the lava. To gather such data, we must land on the red planet and analyze the rocks.

LANDFALL ON MARS: THE *VIKING* MISSIONS

On July 20, 1976, the American probe *Viking 1* achieved the first successful landing on Mars, setting down in the plains of Chryse, in the northern hemisphere. Black-and-white and color imagery unraveled a smooth lava field, rolling out to the horizon, peppered with dark rocks, large boulders and sand dunes several meters high – a landscape reminiscent of eroded lava flows of the American Southwest. The ground showed up pinkish-orange, arguably because of the oxidation of iron-bearing minerals in the rocks and soil – a chemistry suggestive of *mafic* lava, rich in iron and magnesium. Photographed in close-up, many rocks were riddled with small pits that resembled the vacuoles formed in the lava by escaping gases.

No chemical analysis could be performed directly on the rocks.

FIGURE 5.9 Martian lava on *Viking 2*'s landing site, in Utopia Planitia. The central boulder is one meter long and appears banded. The darker and more vesicular left side might be the top of the lava flow, which underwent the most gas exsolution and frothing. Credit: NASA/JPL.

Viking was only capable of testing the soil, primarily to search for microbial life forms – the grand purpose of the mission. The search for life on Mars yielded ambiguous results: no organic molecules were detected, but the soil responded actively when it was supplied with life-sustaining nutrients. In one experiment, a steady emanation of carbon dioxide was observed. In another, oxygen was exhaled, suggesting the presence of mysterious *superoxides* in the soil – metals linked to excess oxygen. Pooling all the data together, project leaders concluded that the soil was responding chemically rather than biologically to the supply of nutrients, although a consensus was never reached.

Whereas biologists controlled four experiments aboard *Viking*, only one experiment was earmarked for geologists: a pocket-sized X-ray spectrometer that could measure the concentration of various atoms in the soil. Silica ranked first among the measured oxides (44%), followed by oxides of iron (17.5%), aluminum (7%), magnesium (6%) and calcium (6%). Sulfur oxide also reached 7%, a much higher figure than in terrestrial soil. Minor elements included titanium oxide (0.5%) and chlorine (0.4 to 0.8%). The greatest surprise, besides the abundant sulfur, came from the very low potassium content, lower than the detection threshold of the instrument (under 0.1%). *Viking 2*, which landed on August 7, 1976, in the northern plains of Utopia, found similar results, hinting that the reddish dust was a well-mixed global average of eroded minerals, distributed by the wind on a planetary scale.

The abundance of iron oxide and the relatively modest amount of silica in the Martian soil indicate that the soil is derived from mafic basaltic rock. As for the vanishingly low level of potassium, it indicates that the magma did not undergo much *differentiation* – or "distillation" – between the time it left the mantle and the time it erupted at the surface. Indeed, when a rising batch of magma precipitates crystals on its way to the surface, its composition changes: the distilled liquid becomes relatively enriched in potassium and other alkali. The fact that the Martian soil is not enriched in potassium proves that most of the magma rose directly to the surface, without changing composition en route. We will later see that on a local scale, some differentiation did take place.

MARTIAN MINERALS

Viking's X-ray spectrometer was able to measure atomic concentrations in the soil, but it stopped short of identifying the minerals that these atoms make up. A few clues were provided by the physical properties of the soil. Close-up imagery of the trenches dug by the sampling arm showed the Martian soil to be flaky and adhesive, two properties of clay minerals. Magnets mounted on the *Viking* probes attracted airborne dust, which likely was a hydrated form of iron oxide belonging to the *maghemite* family (γ-Fe_2O_3).

Scientists further constrained the mineral types present in the soil by "playing scrabble" with the X-ray data. They made up mineral "words" with the atomic "letters" identified by the spectrometer, in such proportions as to use up all the atoms. Some teams mixed selected minerals in the laboratory and adjusted their proportions until they reproduced the *Viking* data. Others ran mineral equations through computers and likewise attempted to match the atomic composition of the Martian soil.

In the lab, the mineral "cocktail" that offered the closest match to the *Viking* data consisted of 47% *nontronite* clay (which includes maghemite as a sub-species), 17% *montmorillonite* and 15% *saponite* (both clays), 13% *kieserite* (a sulfate accounting for the sulfur) and 7% *calcite* (a calcium carbonate). Computer runs also yielded the same ferromagnesian clays: nontronite, montmorillonite and saponite.

These are all familiar minerals. Ferromagnesian clays are widespread on Earth and derive from the alteration of mafic lava in humid environments. As for kieserite and calcite, they precipitate from evaporating brines or crystallize in fissures when groundwater leaches to the surface.

But these minerals, one should recall, are only educated guesses for the composition of the Martian soil. Besides, they are weathering

FIGURE 5.10 Apollinaris Patera.

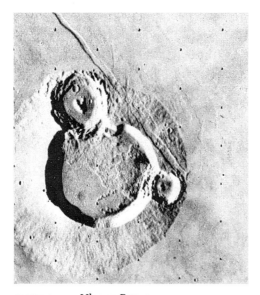

FIGURE 5.11 Ulysses Patera.

Four Martian volcanoes. Apollinaris Patera (Fig. 5.10), Ulysses Patera
(Fig. 5.11), Uranius Tholus (Fig. 5.12) and Tharsis Tholus (Fig. 5.13). See
Fig. 5.4 for locations of last three. In all frames, north is at top.
Apollinaris Patera (Fig. 5.10) is a shield volcano close to the equator,
with a lava-filled caldera and an apron of lava that fans out to the south,
carved by channels. Ulysses Patera (Fig. 5.11), in western Tharsis,
boasts a large caldera and two impact craters that stamp its upper
slopes. The northeastern half of the caldera is covered with hummocky
deposits, while the other half boasts a ring of small dots that might be
volcanic mounds. Uranius Tholus (Fig. 5.12) is an ancient stubby shield
in the northern reaches of Tharsis. It boasts a central caldera, as well as

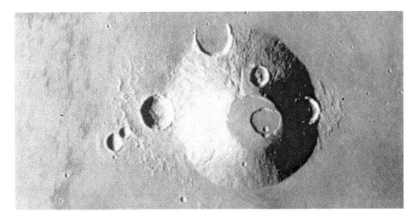

FIGURE 5.12 Uranius Tholus.

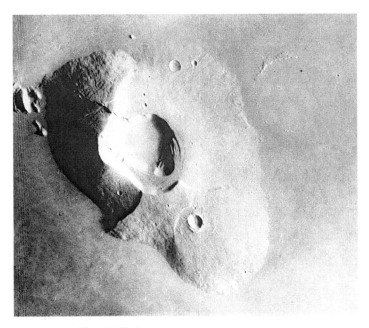

FIGURE 5.13 Tharsis Tholus.

the outline of a previous summit depression. Impact craters gouge its flanks and point to an age in excess of 3 billion years. Tharsis Tholus (Fig. 5.13) lies on the eastern margin of the volcanic province. Its heart-shaped caldera tops a steep structure, 170 km × 110 km, which seems to have experienced a series of flank collapses. The base of the volcano is buried by younger lava flows. Credit: NASA/JPL (Dr. Michael H. Carr, *Viking* Experiment Team Leader), courtesy of the National Space Science Data Center, through the World Data Center-A for Rockets and Satellites (Figs. 5.10, 5.11 and 5.12); and NASA/JPL, courtesy of Brown University (Fig. 5.13).

products, as are most minerals in soil, and say little about the composition of the original lava that volcanologists are so intent on pinning down. Twenty years were to pass before another landing took place on Mars and an X-ray spectrometer gave the first "hard rock" answers.

PATHFINDER AND THE ANDESITE MYSTERY

Mars Pathfinder landed on the southern margin of Chryse Planitia on July 4, 1997. The mission was mostly a technical one – experimenting a new landing system using airbags – but a small rover was carried along with a scientific payload: an alpha/X-ray spectrometer to study the rocks.

The landing site turned out as expected to be a flood plain, littered with rocks and boulders that tumbled down the course of Ares Vallis, a channel flowing out of the highlands to the south. To the west, two low hills – known as Twin Peaks – rose 50 meters above the plains and bore the marks of the flood that swept through the area over two billion years ago. Two days after *Pathfinder* landed, its rover *Sojourner* drove off the ramp and crawled to its first target – a vesicular, foot-long rock baptized "Barnacle Bill" by the science team. The rover positioned its spectrometer against the rock and bombarded a square centimeter of its surface with X-rays and alpha rays, collecting a radiation echo that betrayed its constituent atoms.

The results came as a surprise. With as much as 55% silica, 13% iron oxide and 12% alumina, Barnacle Bill was not truly a basalt – the most common lava on Earth – but a slightly more siliceous igneous rock, known as a basaltic andesite. On Earth, andesites are mainly produced by water-rich volcanism, typical of subduction zones. Less often, they derive from the distillation processes at work in magma chambers – a phenomenon observed in Iceland, Hawaii and the Galapagos Islands.

Coming off this first surprise, *Sojourner* crawled to its next target, "Yogi," a boulder five meters away, shaped like the head of a bear. This time, initial readings showed less silica but a surprisingly

FIGURE 5.14 *Pathfinder's Sojourner* rover cuddles up to the rock "Barnacle Bill" during its 1997 survey of Ares Vallis. The probe's alpha ray spectrometer indicated a chemical composition matching that of basaltic andesite, a type of lava slightly more siliceous than basalt. Credit: NASA/JPL/Caltech, courtesy of the Lunar and Planetary Institute.

high level of sulfur, which led the science team to reconsider their measurements. It appeared that the rock face analyzed by the spectrometer was coated with dust – that same iron and sulfur-rich soil that the *Viking* probes had analyzed years before. By running the Yogi data through a correction algorithm – subtracting the contribution of the dust – scientists came up with new numbers that spelled out basaltic andesite for the second time in a row. Six other tests confirmed the andesitic nature of the rocks.

Again, short of identifying minerals rather than atoms, scientists could only make educated guesses as to the true identity of the Martian rocks. They needed more data and turned to other spacecraft for help. Indeed, just as *Pathfinder* was winding down its work, a new probe by the name of *Mars Global Surveyor* arrived in Martian orbit, on September 11, 1997.

MARS GLOBAL SURVEYOR

Besides a new telephoto camera, capable of picking up details a few meters across on the ground, *Mars Global Surveyor* was equipped with a magnetometer and a remote-sensing instrument called TES: the Thermal Emission Spectrometer.

The spectrometer operated in the infrared. When they are exposed to sunlight, minerals heat up and emit infrared radiation in ways that betray their identity. Their characteristic emission pattern, as a function of wavelength, is known as their *spectral signature*. The TES instrument collected such spectra below the flight line of the satellite – spectra that were then compared in the lab with those of standard minerals. From this matching game, it appeared that the northern plains of Mars, including the landing sites of *Viking* and *Pathfinder*, were rich in plagioclase feldspar, pyroxene and potassium-rich glass, a mineral assemblage consistent with andesite lava. Thus, remote-sensing survey from orbit confirmed the andesite readings that *Pathfinder*'s rover collected on the ground. Another possibility is that the rock is not andesite, but altered, weathered basalt, which would give off a similiar spectral signature.

Less surprising were the infrared spectra collected over the southern hemisphere of Mars. There, the signal corresponded to a mixture of plagioclase feldspar (65%) and clinopyroxene (35%), a mineral make-up typical of basalt. The signature of olivine (a constituent of some types of basalt) was also picked up by the spectrometer over several areas of Mars. It showed up in the layered walls of Valles Marineris, as well as on the bottom of the canyon.

Other patches of olivine-rich material were discovered around the margin of the Isidis impact basin, down slope from the Syrtis Major shield volcano. Careful analysis of the data showed a range of compositions of the greenish-brown mineral, some magnesium-rich and some containing more iron. The presence of olivine around Isidis is probably due to faulting, which exposed outcrops of lava on the rim of the basin.

Infrared remote-sensing only detects what is present in the top soil. It does not reveal most of the lava that covers the Martian surface

PLATE I A gas-rich hawaiian eruption at Pu'u O'o vent, Kilauea eastern rift zone. The hot pyroclasts fall back to fuse together into a lava flow. Credit: J. D. Griggs, U. S. Geological Survey.

The original colour version of plates I to XXIV are available for download at www.cambridge.org/9780521008631

PLATE II A lava flow reaches the sea and explodes into a spray of incandescent fragments and water vapor (Kilauea eastern rift zone). Credit: photo by the author.

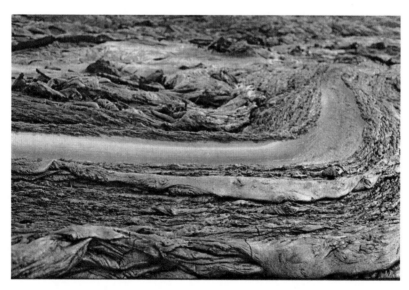

PLATE III Lava flow (Pu'u O'o lava field, Hawaii). The molten basalt builds up levees of solidified rock along the edges of the channel. Credit: photo by the author.

PLATE IV A sector collapse opened up the northern flank of Mount
Saint Helens during the May 18, 1980, eruption. A dome now grows in
the caldera and emits a plume of water vapor. Credit: L. Topinka, U.S.
Geological Survey, Cascades Volcano Observatory.

PLATE V Satellite view of Cerro Galan resurgent caldera, Argentina.
The caldera, in the shape of a pear, measures 35 by 20 km. Dark lava
domes are noticeable on the periphery: west of the lake, right of frame,
and top of frame. Credit: NASA/*Landsat*, color-composed by Peter W.
Francis at the image processing facility of the Lunar and Planetary
Laboratory, Houston. Courtesy of Peter W. Francis from his book
Volcanoes of the Central Andes, Springer-Verlag, 1992.

PLATE VI A volcano in Kamchatka (Russia) erupts a thick plume of ash that blows out over the Pacific. Volcanoes contribute large amounts of volatiles to the Earth's atmosphere, and their ash clouds affect the climate. Photo by the crew of the Space Shuttle (STS-68, 1994). Credit: NASA/JSC, courtesy of Debra Dodds.

PLATE VII False-color view of the Moon, imaged by the *Galileo* spacecraft on its way to Jupiter. Titanium-rich soils show up blue in the Sea of Tranquillity (left), whereas titanium-poor soils in the Sea of Serenity show up orange (center right). Credit: NASA/JPL-Caltech.

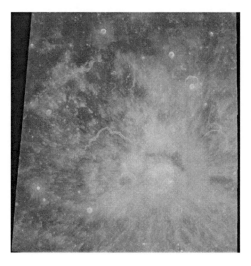

PLATE VIII False-color view of the Aristarchus plateau on the Moon, imaged by the *Clementine* orbiter. Iron-rich dark mantling deposits, believed to be volcanic ash, show up deep red. Iron-poor highland material is blue. Credit: NASA/U. S. Geological Survey

PLATE IX Astronaut Harrison Schmitt (*Apollo 17*) on the rim of Shorty crater (right) in the valley of Taurus-Littrow. Subtle streaks of orange volcanic soil are seen at the foot of the mound (top left), and inside the crater (top center and bottom right). Credit: NASA/JSC.

PLATE X Close-up of the orange volcanic soil, on the rim of Shorty crater (valley of Taurus-Littrow, *Apollo 17*). The astronauts dug a trench to sample the soil. Credit: NASA/JSC.

PLATE XI Mars viewed by the MOC camera of *Mars Global Surveyor*. The western half of the image shows the volcano Olympus Mons, shrouded by clouds, on the left margin. The cloudy patch of Alba Patera is to the upper right of it (halfway to the north polar cap) and the diagonal threesome of the Tharsis Montes to the lower right. Credit: NASA/JPL-Caltech/Malin Space Science Systems.

PLATE XII Mars: Olympus Mons, the tallest volcano in the Solar System, is topped by a complex caldera and crowned by cirrus clouds. *Viking* image. Credit: NASA/JPL.

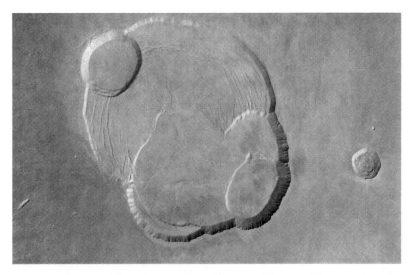

PLATE XIII Mars: the complex caldera atop Olympus Mons consists of six nested pits that collapsed in turn, as magma shifted below the summit. Credit: ESA/DLR/FU Berlin (G. Neukum).

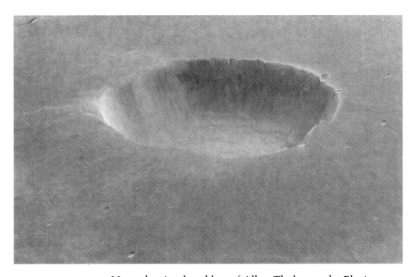

PLATE XIV Mars: the simple caldera of Albor Tholus on the Elysium plateau is over two kilometers deep. A sheet of dust cascades down the western side. Credit: ESA/DLR/FU Berlin (G. Neukum).

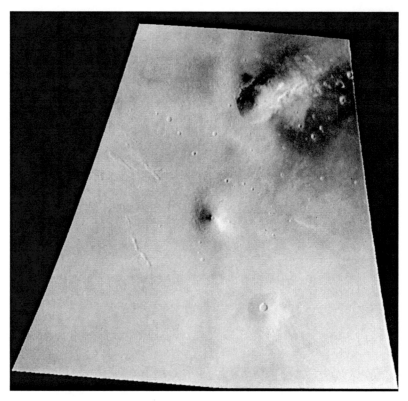

PLATE XV Mars: The Elysium province boasts three large volcanoes. From top to bottom: Hecates Tholus, Elysium Mons and Albor Tholus. Credit: NASA/MGS/Malin Space Science Systems.

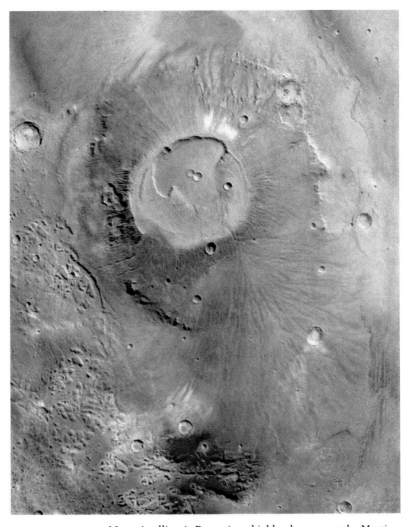

PLATE XVI Mars: Apollinaris Patera is a shield volcano near the Martian equator, with a 60-km-wide caldera and a fan of lava flows that flow down the southern flank. *Viking* image. Credit: NASA/JPL-Caltech.

PLATE XVII Venus and Earth compared. Venus is slightly smaller than the Earth. Its topography is dominated by plains, with few deep basins (dark blue). Uplands (yellow) comprise Ishtar Terra (north) and rift zones and volcanoes along the equator (Aphrodite Terra). Credit: NASA/JPL-Caltech, courtesy of Brown University.

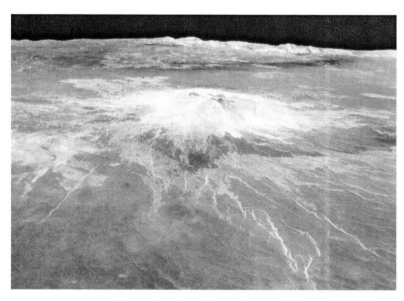

PLATE XVIII Perspective view of Sapas Mons, a 600-km-wide shield volcano on Venus. View is to the north. In this radar image, brightness usually indicates roughness of the lava flows. Credit: NASA/JPL-Caltech, courtesy of Brown University.

PLATE XIX The ground of Venus, imaged by the Russian probe *Venera 13* in 1982. The slabs of rock are believed to be alkaline basalt, based on the results from the probe's spectrometer. Credit: USSR Academy of Science, courtesy of the Vernadsky Institute and Brown University.

PLATE XX Two volcanoes rise on the horizon of Venus in this perspective view constructed from *Magellan* radar imagery. Sif Mons is to the left and Gula Mons to the right, with a fracture belt in the foreground. Credit: NASA/JPL-Caltech, courtesy of Brown University.

PLATE XXI *Galileo* image of Io, covered with lava flows (dark) and sulfurous deposits (light). The eruption of volcano Pillan is observed on the limb, with a 120-km-tall plume. In center frame, the circular spot with the dark hooked-shaped marking is Prometheus volcano.
Credit: NASA/JPL-Caltech.

PLATE XXII The eruption of Masubi volcano is caught by *Galileo* on the limb of Io. Credit: NASA/JPL-Caltech.

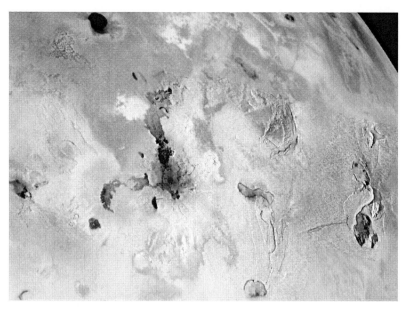

PLATE XXIII The eagle-shaped Amirani-Maui lava field dominates this landscape of Io, imaged by the *Galileo* spacecraft. Credit: NASA/JPL-Caltech.

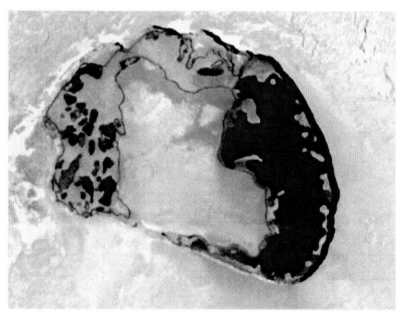

PLATE XXIV Tupan Patera imaged by *Galileo* during a low-altitude pass over Io. Dark patches are recent lava. Orange and greenish patches are older lava covered by sulfurous deposits. Credit: NASA/JPL-Caltech.

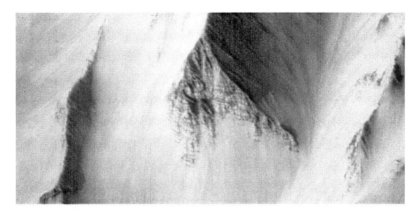

FIGURE 5.15 The precipitous walls of Valles Marineris display a thick stack of horizontal layers. These beds are voluminous lava flows that erupted during the first half of Martian history.
Credit: NASA/JPL/Malin Space Science Systems.

but is hidden under several centimeters of weathered soil and dust. Instead, it will pick up the signature of this weathered material, which can also be enlightening. One of the great discoveries of TES happens to be patches of hematite, an iron-rich oxide (α-Fe_2O_3) that is one of the weathering products of basalt. In fact, Mars owes its red color to its abundant hematite and other iron-rich dust – specks a few micrometers in size. But besides this "background" hematite, TES picked up an unusually strong signal in two areas of Mars: one wide patch, the size of Ireland ($500\,km \times 200\,km$), in the equatorial region of Terra Meridiani ($0°\,N$, $0°\,W$), and a smaller patch in an old impact basin nearby, known as Aram Chaos ($2°\,N$, $21°\,W$).

The strong hematite signal in these areas indicates unusually large mineral grains, of the kind that form in bodies of standing water – a clue that Mars was once covered by lakes or shallow seas. Leached out of the lava seabed, the iron precipitated to form a tell-tale sedimentary layer.

SPIRIT **AND** *OPPORTUNITY*
Ancient seas on Mars would constitute an environment conducive to the emergence of life, especially if hydrothermal waters from volcanoes

provided the necessary heat and dissolved nutrients (in the form of metallic ions). A new generation of landing probes was assigned the task of checking out on the ground the promising sites identified from orbit.

Launched during the summer of 2003, two exploration rovers named *Spirit* and *Opportunity* landed on two such sites in January of 2004: Gusev crater – an impact basin that stood in the pathway of an ancient flood channel – and the hematite-rich area of Terra Meridiani.

To the delight of the science team, Terra Meridiani turned out as expected to be an ancient lakebed. Delivered in its airbag cocoon, *Opportunity* rolled to a stop in the middle of a small impact crater that exposed tantalizing layers of bedrock. The sophisticated rover, equipped with camera, microscope and three different spectrometers, only had to drive a few meters to analyze the light-colored outcrop: this was no lava, but indeed lakebed sediments, full of little beads of hematite, christened "blueberries" by the science team.

The sediment itself turned out to be mainly composed of sulfates: sulfur compounds rich in calcium, magnesium, potassium and sodium. The abundance of sulfur pointed to strong volcanic activity at the time the outcrops formed, close to four billion years ago. The volcanic gas dissolved in the water to form diluted sulfuric acid that ate away at the underlying lava beds, leached out the metals, and turned them into sulfate salts that piled up on the lake bottom. Such an environment might well have harbored biological activity, and scientists are already dreaming of sending a sample-return probe to Terra Meridiani in the future, to bring back such sulfates to Earth for further analysis.

Opportunity's twin probe *Spirit* also performed a flawless landing in Gusev crater, halfway around the planet. From orbit, Gusev shows layered delta-like terrain that likewise points to lakebed sediments, but the rover did not come to rest within immediate reach of such a tell-tale outcrop. To the delight of volcanologists, however, the rolling plains were littered with blocks of lava that could be analyzed by the rover with more sophisticated instruments than the one carried by *Sojourner* back in 1997.

The new rover carried a rock abrasion tool to clear the rind of

FIGURE 5.16 *Spirit*, one of the twin Mars exploration rovers (2004), discovered blocks of lava, sculpted by the wind, on the floor of Gusev crater. This rock, nicknamed "Humphrey," was brushed off by the rover's robotic arm (dark circles) before analysis. The results indicate that Humphrey is an olivine basalt. Credit: NASA/JPL/Cornell/USGS.

FIGURE 5.17 Close-up of the rock "Humphrey," taken by the microscopic imager on board the *Spirit* rover. The rock surface, polished by an abrasion tool, shows bright specks (holes) where mineral grains were plucked out, and darker patches that are crystals of olivine. Humphrey is an olivine basalt that contains approximately 30% olivine and 40% plagioclase feldspar. Credit: NASA/JPL/Cornell/USGS.

weathered material off the lava and reveal a clear rock face for analysis. The first rocks analyzed by *Spirit*, named Adirondack and Humphrey, are unquestionably basalt. Whereas *Pathfinder* had found the lava in the northern plains to be silica-rich – close to andesite in composition – the lava in the equatorial region of Gusev is silica-poor. Its chemistry indicates that it contains approximately 30% olivine, 40% plagioclase feldspar and 20% pyroxene. It is therefore an olivine basalt, and close-up imagery taken with the rover's microscope indeed shows the dark, six-sided crystals of olivine on the polished rock face.

Where did the lava come from? Did it erupt directly from under Gusev crater, through the fractures that lace the floor of the impact basin? Did it flow in from outside the basin, perhaps from the nearby Apollinaris Patera volcano to the north? Or were the lava boulders carried in by the flood channel that empties into Gusev crater from the south? It will be for future Martionauts to find out.

METEORITES FROM MARS

One can only guess so much about Martian lava from robotic probes that carry only a few kilograms of scientific equipment. What geologists truly want is to examine these rocks on Earth, with the full power of their laboratory equipment. However, sample-return missions, whereby robotic probes will bring back to Earth a small capsule-worth of Martian rock and soil, are not expected before 2014 at the earliest. And even then, mission planners will probably wish to bring back sedimentary rock rather than lava.

This is where volcanologists get some unexpected help from the heavens. It appears indeed that several dozen Martian rocks have been delivered, free of charge, to our doorstep. These are the so-called "SNC" or Martian meteorites.

On October 3, 1815, at 8 o'clock in the morning, the farmers of Chassigny, a small village of eastern France, heard a powerful blast and sighted a trail of smoke in the sky. They retrieved a large meteorite on the ground and sent it to the Academy of Science in Paris.

FIGURE 5.18 A rock from Mars? The Chassigny meteorite fell to Earth in 1815. Its unusual chemistry and its relatively young age (1.3 billion years) point to a Martian origin. The rock is a cumulate of olivine crystals (dunite) that formed at the bottom of a magma chamber. Credit: courtesy of the Laboratoire de Minéralogie, Museum National d'Histoire Naturelle, Paris.

The rock from space had an unusual mineralogy: it was composed entirely of olivine crystals. Dating techniques would later prove it to be 1.3 billion years old, much younger than ordinary meteorites, which all date back to the birth of the Solar System, 4.5 billion years ago.

Over the years, more of these young unusual meteorites were retrieved at Shergotty in India, Nakhla in Egypt, Indiana and California, Brazil, Nigeria, Antarctica, Morocco . . . Known collectively as the "SNC" meteorites – the initials of the first three specimens – their numbers had grown to eleven samples before their Martian origin was suspected in the early 1980s. Their young age could only be explained if they were forged within a large planet that had stayed hot and active up until recently. Their oxidized state also pointed to a planetary cradle, rather than to the oxygen-poor asteroid belt. The final clue, that tipped the balance, was the discovery in the eleventh specimen of the family – ALH 79001 from Antarctica – of small bubbles of gas that were analyzed in the lab. The extracted carbon dioxide, argon and other gases displayed proportions and isotopic ratios that were characteristic of the Martian atmosphere, as measured *in situ* by the *Viking* probes. They were rocks from Mars!

Scientists soon came up with a viable mechanism to explain the rocks' transfer from Mars to Earth. Apparently they were ejected from the red planet by asteroid and comet impacts. Many SNC meteorites contain shocked minerals – proof of their violent ejection – and all bear the mark of cosmic rays, which indicates that they drifted in the vacuum of space for millions of years before falling to Earth.

The twelfth Martian meteorite was discovered in Antarctica in 1984. Labeled ALH 84001, it became famous twelve years later when a research team announced that it contained fossilized bacteria from Mars. As we shall later see, this theory is widely disputed, but it certainly brought Martian meteorites into the limelight. New specimens are periodically discovered. By 2004, over 40 finds have been reported, belonging to at least 28 different rocks, and the harvest is not about to end. Besides the scores of Martian meteorites lying undiscovered on the continents and on the ocean floors, several more fall to Earth each year.

Martian meteorites are a bonanza for volcanologists. All specimens so far are igneous rocks. Sedimentary rocks might also fly in from the red planet, but they are harder to single out amidst terrestrial rocks and have so far escaped the scrutiny of meteorite hunters.

The igneous meteorites from Mars are divided into fine-grained volcanic specimens, akin to basalts, which crystallized near the surface of the red planet, and larger-grained *cumulates* that formed deeper underground, as crystals settled to the bottom of magma chambers.

The former, basaltic meteorites are known as *shergottites*, in reference to the first such specimen discovered at Shergotty, India, in 1865. As of 2004, a dozen other shergottites have been discovered in Nigeria, Antarctica, California, Oman and northwest Africa, and the list is bound to grow. They all contain *clinopyroxene* and *plagioclase* crystals – the characteristic mineral assemblage of basalts.

Subtle variations between shergottites point to different magmatic histories. For example, the California finds, known as the Los Angeles meteorites, are magnesium-poor. Their parent magma appears to have lost a sizeable fraction of this metal and of other elements that were scavenged by crystals sinking to the bottom of the magma chamber. This fractional crystallization led to a "distilled" batch of magma that later erupted to the surface to yield magnesium-poor basalt.

The other class of Martian meteorites is the counterpart of these distilled basalts, i.e. the dregs or cumulates of crystals that were left behind at the bottom of magma chambers. The Chassigny meteorite is a cumulate of olivine crystals, known as dunite. Likewise, the Nakhla meteorite and half a dozen other specimens are cumulates of clinopyroxene crystals, and the ALH 84001 meteorite is a cumulate of orthopyroxene.

Finally, four specimens come from deeper in the Martian crust and represent the "source rock" that melts to yield magma. Known as *peridotite*, these samples consist of a mosaic of olivine, pyroxene and plagioclase crystals.

The first conclusion one can draw from this line-up of Martian suspects is that the vast majority belong to the basalt family – classic basalts and cumulate dregs from basaltic magma. This is somewhat of a contradiction, since half of the Martian lavas identified from orbit are supposed to be andesites. Why do we get only basalts? Why is there not a single andesite among the heist of meteorites from Mars?

One answer is that the meteorites are not representative of the planet as a whole, despite their apparently large numbers. In other words, the 40-odd specimens do not come from 40 different sites on Mars. Rather, each asteroid or comet impact launched a salvo of rocks into space. We must then regroup the meteorites into a smaller number of impact events. This sorting is possible because their exposure time to cosmic rays records when the meteorites were launched

into space. For instance, the Nakhla, Lafayette and Chassigny meteorites were launched off Mars 10 million years ago, allegedly by the
same impact from the same site. Zagami and Shergotty were blown
off from another site, 2.6 million years ago. The bottom line is that
the SNC meteorites collected so far come from only half a dozen locations on Mars. The fact that there are no andesites among the SNCs
hints that the so-called andesite detected on Mars by remote-sensing
might rather be a form of altered basalt.

YOUNG LAVA

Since we can date the Martian rocks in our possession, we get some
direct insight into the duration of igneous activity on the red planet.
One specimen dates back to the first days of Martian history: ALH
84001 is 4.4 billion years old. Next comes the Chassigny-Nakhla-
Lafayette group with a jump in time to 1.3 billion years. The Dar Al
Gani meteorite appears to be 475 million years old, and another specimen 327 million years old. As for Zagami, Shergotty and the whole
sleuth of shergottite basalts, they claim an age of 175 million years –
very young lavas indeed.

These ages support the recent view that Martian volcanism did
not shut down billions of years ago, and that some volcanoes in the
Tharsis and Elysium regions are a few hundred million years old at
most. We do not know from what locations on Mars each set of meteorites was ejected, but we can make educated guesses by matching
rock ages with postulated surface ages, as determined by crater counts.
For instance, there is little doubt that the 4.4-billion-year-old ALH
84001 originated from the heavily-cratered highlands. Chassigny and
the Nakhlites (1.3 billion years) would come from a moderately-
cratered area, such as the Elysium plateau or the older Tharsis volcanoes. As for the Shergottites (175 million years), they can only come
from the sparsely-cratered Elysium basin or the flanks of the youngest
Tharsis volcanoes.

If we want to be bolder yet, we can take bets as to the exact
source locations of the Martian meteorites. We know that they were

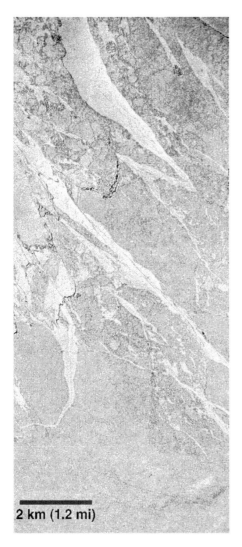

2 km (1.2 mi)

FIGURE 5.19 This very young lava field in the Elysium basin shows rafted plates of lava that split apart during their emplacement. They are separated by wedges of smoother material. The lack of impact craters points to an age of a few tens of millions of years at most. Credit: NASA/JPL/Malin Space Science Systems.

ejected less than ten million years ago, according to their cosmic ray exposure. The search thus boils down to volcanic areas on Mars struck by young impact craters of that age. The latter should be 10 km or more in diameter and preferentially oblong, since large, oblique impacts are the most likely to eject rocks into space.

One young oval impact crater impinges on the lower flank of Ceraunius Tholus volcano in northern Tharsis. The volcano looks

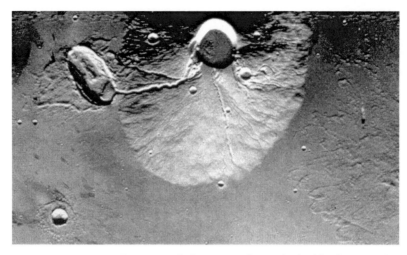

FIGURE 5.20 Ceraunius Tholus is a 100-km-wide shield volcano on the
Tharsis plateau. A lava channel descends from its caldera to an oval
impact crater at the foot of the slope. Such an impact might have flung
middle-aged lava rocks from Ceraunius Tholus towards Earth and could
be the source of the 1.3 billion-year-old Chassigny and Shergotty
Martian meteorites. Credit: NASA/JPL (*Viking*), courtesy of Brown
University.

like it could be 1.3 billion years old (the age of the Nakhla meteorite
group), and the impact hit close enough to the magma chamber at the
base of the edifice to eject cumulate-type rocks such as the Chassigny
dunite and the Nakhla pyroxenites.

As for the Shergottite meteorites, the leading candidate is
Olympus Mons, where three impact craters are under scrutiny: one
on the young lava plains surrounding the volcano and the other two
on the upper flanks of the edifice, near the caldera. A second possible
source area is the Cerberus lava plains, south of the Elysium volcanoes. These flows have a very fluid morphology, typical of basaltic
lava, with large plates of solidified crust that were rafted along by the
fiery current. Impact craters are few and testify to the young age of
the Cerberus flows; half a dozen of these fresh-looking craters qualify
as launching platforms for the Shergottites.

In conclusion, meteorites from the red planet inform us about the make-up and age of the Martian lavas. But they also tell us something about the environmental conditions that they experienced near the surface of Mars. Indeed, any fluids circulating through the rocks would deposit "secondary" minerals in their cracks and fractures. These minerals have been recognized in some of the meteorites and shed light on the hydrothermal – and arguably biological – processes at work on the red planet.

HYDROTHERMAL ACTIVITY AND LIFE ON MARS

The most famous "secondary" minerals discovered in a Martian meteorite are the carbonate globules lining the fractures of Antarctic specimen ALH 84001. A quarter of a millimeter in size, these orange grains of calcium, iron and magnesium carbonate have become world-famous, since a team of investigators announced in 1996 that they might contain evidence of ancient Martian life.

The 1976 *Viking* missions, we must recall, did not find extant life in the surface soil of Mars, although the results were far from conclusive. But however negative, the results do not rule out the possibility that life might have existed in the past, or even that it still exists today in the warmer environment of underground rocks, protected from the cold dryness and the ultraviolet radiation that sterilize the surface.

The carbonate globules of ALH 84001 are promising in this respect. They were deposited in a rock that was buried tens of meters deep (according to the size of its crystals) and that dates back to the early days of Martian history. The rock is 4.4 billion years old, and the carbonates are dated at 3.9 billion years (they formed half a billion years after the rock itself). Several lines of evidence suggest that the carbonate deposits involved biological processes. Minuscule features, imaged with an electronic microscope, resemble fossil bacteria; magnetite crystals imbedded in the carbonates resemble those secreted on Earth by certain strands of bacteria; and organic material including PAHs (polycyclic aromatic hydrocarbons) are detectable by mass spectroscopy.

FIGURE 5.21 The Martian meteorite ALH 84001 was found in Antarctica in 1984. A cumulate rock of orthopyroxene crystals, it probably formed at the bottom of a magma chamber. Its fractures host carbonate crystals, deposited by carbon-rich fluids, that bear fossil-like features. Credit: NASA/JSC.

Critics have challenged both the Martian origin and the biological nature of these clues. They explain them instead by inorganic processes and contamination of the meteorite by terrestrial bacteria since it fell to Earth.

It is also far from certain that the carbonate globules indicate pervasive circulation of water in the rock fractures. ALH 84001 contains few water-bearing minerals that would attest to such a wet environment: it appears that the rock was only briefly exposed to fluid, be it water or carbon dioxide.

Other Martian meteorites are more encouraging about the prospects of hydrothermal activity on Mars – and therefore about the potential for underground biological activity. The Nakhlites contain water-bearing minerals, such as clays and hydrated iron oxides. One Nakhlite in particular, the Lafayette meteorite, contains substantial amounts of *iddingsite* – a rust-colored mixture of iron hydroxides.

FIGURE 5.22 This electron microscope close-up of a carbonate spherule, inside the Martian meteorite ALH 84001, shows a segmented rod that resembles a microfossil and could be evidence of ancient life on Mars. Alternately, it could be the fossil of a terrestrial bacterium, since the meteorite lay on the Antarctic sheet for more than 15000 years before being collected. Credit: NASA/JSC.

Lafayette, like all Nakhlites, crystallized at the bottom of a magma chamber (or lava pool) 1.3 billion years ago, but the iddingsite minerals have been dated at 670 million years (at most), proof that a pulse of water briefly ran through the underground rock layers at that time. Volcanic activity might then have provided favorable conditions for life – liquid water and soluble ions – in select locations and during select time frames of Martian history.

On Earth, microbial life thrives in hydrothermal systems. Algal and bacterial mats grow in hot water pools at Yellowstone, in Iceland, New Zealand and other volcanic centers. Bacteria also bloom around hydrothermal vents on the deep ocean floor, where they derive their energy not from the Sun, nor from the recycling of dead organic material, but directly from the transformation of dissolved chemicals. Some bacteria catch the sulfate ions in the water and turn them into sulfides – harvesting electrons in the process – while others pump the dissolved hydrogen and carbon dioxide out of the water and derive their energy by combining the pair into methane gas.

Microbial life does not only thrive in open water. Bacteria are now discovered deep underground in the tiny fractures of rocks. Cores retrieved 3000 meters below the surface, notably in the basalt layers of the Columbia River Plateau, contain microorganisms that are a fraction of a micrometer in size, known as *nanobacteria*. These primitive organisms, which also derive their energy from the reaction of hydrogen and carbon dioxide, have a very slow metabolism since the flow of water and nutrients is very slow through the fractures. Instead of dividing into daughter cells every half hour or so – as bacteria do at the surface when nutrients are plentiful – underground nanobacteria reproduce only once or twice per century. Such slow metabolism and reproduction rates make these life forms very difficult to detect.

If life did emerge on Mars – autonomously or seeded by meteorites from Earth – microbial life forms might similarly live in underground rocks, where heat, liquid water and dissolved minerals and gases are available, and where the harsh surface conditions are avoided. Proving their existence, however, will be difficult. One way of establishing the presence of life is to detect gases in the atmosphere, which are out of chemical equilibrium with the environment and can only be replenished by biological activity. On Earth, methane is such a tell-tale gas. However, an underground biosphere limited to low-metabolism bacteria might not convey enough gas to the surface to betray its existence.

A more promising way to detect Martian life is to discover its fossil remains, amassed over many generations. In the case of sulfate-reducing bacteria, these remains might be in the form of sulfide deposits – pyrite crystals and the like. Some of these underground fossil beds could then be uncovered by erosion or blasted to the surface by meteoritic impacts.

The search is on. Besides *Mars Global Surveyor* that established a crude map of chemical and mineralogical provinces on Mars, the

American orbiter *Mars Odyssey* has begun an infrared mapping of the Martian surface in 2002. It has been joined in orbit by the European probe *Mars Express* in December of 2003, which also carries an infrared spectrometer. This survey of Mars aims to pinpoint the areas where water might have pooled, when the atmosphere was thicker than at present, notably during times of abundant volcanism.

PLATE TECTONICS ON MARS

Volcanic activity is the principal mechanism by which a planet releases its deep-seated gases to the surface and builds itself an atmosphere. Mars was no exception and its volcanoes certainly spewed vast quantities of water and carbon dioxide into the environment. This contribution was greatest in the early days of Martian history, when planetary heat, volcanism and degassing were at their maximum. They have declined ever since. Assessing the exact volume and timing of volcanic degassing should therefore shed light on the climatic history of the red planet.

The early era of Martian history, known as the *Noachian* (roughly 4.5 to 3.5 billion years ago), was characterized by heavy asteroid and comet bombardment and abundant volcanism, although evidence of the latter is partially erased by impact craters and other forms of erosion. However, the pockmarked plateaus of the southern hemisphere, which date back to this era, bear indirect evidence of igneous activity in the form of a "fossil" magnetic field.

Detected by the magnetometer aboard *Mars Global Surveyor*, this magnetization of the crust proves that Mars had a magnetic field early in its history, when it had a molten iron core. This field was recorded by metal-rich magma when it rose through the Martian crust, cooled and solidified.

On Earth, this phenomenon is responsible for the pattern of magnetic stripes that runs through the ocean basins. Because the ocean crust expands by injection of magma along the mid-ocean ridges (in conveyor belt fashion), and because the Earth's magnetic field periodically reverses direction during the process, the magnetization of the ocean basins takes the form of parallel stripes of contrasting intensity.

On Mars, a similar pattern of parallel magnetic anomalies is recognized in the southern plateaus, hinting that the crust also grew by injection of long, dike-like swaths of magma. This vast magnetized region lies in Terra Cimmeria and Terra Sirenum, highland plateaus that straddle the 180°W meridian in Mars' southern hemisphere. Fainter stripes can also be detected in the northern lowlands.

The parallel, magnetized swaths of terrain in the south can be traced for about 2000 km and are close to 200 km wide. There are over a dozen such strips. What is remarkable is the intensity of the "fossil" magnetic field. Flying more than 100 km over the landscape, *Mars Global Surveyor* registered values as high as 1500 nanoteslas, tens of times higher than magnetic anomalies on Earth, when observed from the same altitude.

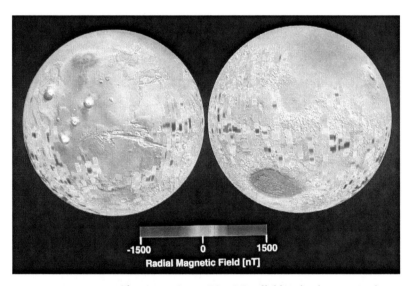

FIGURE 5.23 Plate tectonics on Mars? Parallel bands of magnetized terrain, in the ancient plateaus of the southern hemisphere, can be traced for up to 2000 km. They are reminiscent of seafloor spreading on Earth and could likewise point to an early form of crustal spreading on Mars. Credit: NASA (*Mars Global Surveyor*)/JPL/GSFC, courtesy of Lunar and Planetary Institute.

One can assume that the Martian magnetic field was stronger than the Earth's at the time, but only to a point. The high figures must also mean that the sheets of metal-rich magma extend to great depths in the Martian crust – probably down to 30 km, as opposed to 5 km under the Earth's mid-ocean ridges. Thirdly, minerals highly susceptible to magnetization, such as hematite and pyrrhotite, might also be involved.

Be it as it may, large sheets of magma were injected into the Martian crust early in its history. But if the crust was widening in one area, this means that a primitive form of plate tectonics was at work. In other places the crust would have shrunk, in order for the areas to balance out, and Mars would have been divided into jostling plates with transform faults and subduction zones. Although there is little evidence of such a tectonic layout at any time in Mars' history, some patterns are worth noting.

In one study, the alignment of the three Tharsis volcanoes – Arsia, Pavonis and Ascraeus Montes – might mark the margin of a tectonic plate, where Martian crust was once subducted and sank back into the mantle. In this scenario, the plate shrank faster on one side than it expanded on the other. It got completely swallowed up, until the still rifting and buoyant margin made contact with the trench and jammed the motion. Olympus Mons and Alba Patera would then have grown over the still active, but immobilized rift zone. This bold model, which suggests crustal motion in the northern hemisphere rather late in Martian history, will need to hold up to the scrutiny of future magnetic and tectonic surveys.

VOLCANOES AND THE MARTIAN CLIMATE

There is evidence of flowing water early in Martian history. The old southern plateaus – those very plateaus that show magnetic anomalies – are streaked with ancient river beds. These valley networks are short and immature. The limited number of confluents does not speak of frequent and sustained rainfalls. Instead of precipitation out of the atmosphere, one can imagine that groundwater welled up from below, sapping the surface to create the networks. There is also the possibility that the valleys formed under a sheet of ice, as suggested by the remarkably similar "candelabra" networks that formed under glaciers in the Arctic on Earth. Regardless of their exact mode of formation, the ancient valley networks involved substantial amounts of water, which were released by volcanoes in eruption plumes directly into the atmosphere and through the discharge of hydrothermal groundwater. Carbonic gas was also released in the process. Certain chemical markers, like argon, indeed suggest that a thick blanket of gas once existed, with a pressure of at least a third of a bar (0.03 MPa), and even up to 2 bars (0.2 MPa) according to some specialists.

The northern Tharsis rise also harbored abundant volcanism in the Noachian era (4.5–3.5 billion years). Although the rise is covered by more recent lava, there are ways of estimating the volume of magma that was emplaced at the time. It so happens that Tharsis is carved open by a gigantic rift: Valles Marineris. Its towering cliffs display the pile-up of Noachian lava flows in cross-section. Thanks to the telephoto imagery taken by *Mars Global Surveyor*, volcanic layering is recognizable from top to bottom of the cliff face, over 8000 m of vertical drop! Assuming that this thickness of Noachian lava underpins the entire Tharsis plateau, the total volume reaches hundreds of millions of cubic kilometers! How much gas did this early pulse of volcanism contribute to the Martian atmosphere?

If we assume that the magma contained 0.65% of dissolved carbon dioxide (a figure typical of basaltic melt), then the early Tharsis lava would have released enough CO_2 to build up the atmosphere to a pressure of 1.5 bar – one and a half times the pressure of the Earth's atmosphere! Moreover, if the magma contained 2% of dissolved water (a figure also typical of basaltic melt), an extra bar or two of water vapor would have entered the atmosphere, equivalent – if condensed and rained to the ground – to a global ocean 120 meters deep over the entire planet.

There are many hints that large bodies of water occupied the lowlands of Mars in the past. Laser altimetry indicates extremely flat plains in the north, as flat as the abyssal plains that cover the Earth's ocean floors. Thick sedimentary layers, probably deposited in water, are stacked in the deep basins of Valles Marineris. Giant channels lead from the highlands into the basins and testify to water circulation on a vast scale.

What happened to this thick atmosphere and to the putative oceans? Some of the Martian water is still around today. There is a water ice cap at the north pole that is the size of Greenland and reaches a thickness of 3000 meters. If it was melted and its water spread over the entire surface of Mars, it would result in a global ocean tens of meters deep. More water is stored underground as permafrost in the high latitudes of Mars. But the majority of the water and carbon dioxide released by Martian volcanism was simply lost to space. Weakly bound by the planet's low gravity, gases were stripped away by the solar wind and blasted away by large impacts. Calculations show that the bleeding away of a thick (1.5 bar) atmosphere would occur in a few hundred million years. Thus, when volcanic activity dropped below a certain threshold, the gas input to the atmosphere was no longer able to balance atmospheric losses, and the pressure dropped. This transition from a pressurized and wet planet to a dry vacuum-like desert probably occurred during the Hesperian era (3.5–2.0 billion years ago). Surges of gas and water occurred every time an igneous event released volatiles from the mantle or the crust – as attested by the catastrophic outflow channels that tear up the Martian landscape. During the second half of Martian history – the late Hesperian and Amazonian eras – the rate of igneous activity continued to decline and with it the role of volcanism in sustaining the Martian atmosphere.

The extensive lava plains of the Hesperian period, for example, probably degassed ten times less volatiles than the thick formations of the Noachian. Carbon dioxide might have amounted to 0.2 bar of additional pressure (200 hectopascals), and the layer of condensable water – spread uniformly over the planet – would have reached a depth of only ten meters. Volcanic shields, such as Alba Patera and Olympus Mons, would have contributed even lesser quantities of fluids. Their total output probably amounted to a couple of meters of global water and to 20 hectopascals of carbon dioxide – four times the present atmospheric pressure (5 hectopascals).

Working down the scale, how much would one giant eruption be able to contribute to the Martian atmosphere, if it were to occur today? If we take the upper ash unit of Alba Patera as an example and assume that this unit is 100 m thick, the volatiles released during the eruption would only raise the atmospheric pressure by 5% (from 5 to 5.25 hectopascals) and precipitate only a mere centimeter of liquid water onto the surface of Mars. Summing up the figures, we might infer that volcanic activity played a major role in creating a thick atmosphere and shallow oceans in the first era of Martian history (the Noachian), provided discrete pulses of "refill" in the evanescent environment of the second era (the Hesperian) and only weakly affected the thin dry atmosphere of the third era (the Amazonian).

The reality is probably more complex, since a little cause can trigger a large effect. We have seen on Earth how tricky it is to model climate change. Might a small input of carbon dioxide on Mars set off a mild greenhouse effect that would heat the topsoil, release trapped volatiles, and "snowball" to the point of triggering a major climate upheaval? There are signs that Mars underwent unusual events in recent history – water run-off down steep gullies and even major out-flows along recent-looking channels. How much are these events influenced by astronomical factors – such as the cyclical tilting of Mars' rotation axis and the periodic variations of its orbital eccen-tricity – and how much are they affected by resurgent volcanic activ-ity? Perhaps the ice and dust trapped in the polar caps will shed light on the climatic history of the red planet, as they do here on Earth. The future exploration of Mars will certainly focus on such areas, where the planet's climate and volcanic history can be deciphered.

EXPLORING MARS

Besides the polar caps, the great rift of Valles Marineris is one of the key locations where one would want to land a probe or a manned expedition. The rift is near the equator – which provides good solar illumination and excellent radio contact with Earth – and offers a variety of intriguing landforms to study.

We have already noted the spectacular record of early volcanism expressed in the layered cliffs of Valles Marineris. What lies at the bottom is equally interesting. In many places, stacks of layered sediments rise as steep, tall mesas, sometimes reaching the level of the canyon rim. Typically, they are separated from the canyon walls by deep moats. It is tempting to interpret these deposits as being marine or lacustrine sediments, but where did they come from? Why do they stop short of the canyon walls? How come they are so thick and rise so high?

Geologists have suggested that these mesas are volcanic *tuyas* – volcanoes made of pillow lavas and ash layers, which form under a cover of ice. On Earth, the showcase examples are in Iceland. Under ice, the magma is confined by the pressure, its volatiles stay in solution, and only smooth pillow lavas are erupted. As the edifice builds up – melting its ice coffin in the process – the overhead pressure is lowered and the volatile-rich magma erupts explosively, covering the lava mound with steeply-inclined beds of ash. If the edifice finally emerges from the ice and the melt water, a hard cap of subaerial lava might crown the complex.

These characteristics – steep profile, sloping beds and resistant cap rock – are precisely those recognized in the strange mounds of Valles Marineris. Volcanoes on the floor of Valles Marineris would not be surprising, since we can expect magma to erupt on the floor of a rift, rising to the surface along its bounding faults. On Earth, the Great African Rift and the basins of the American Southwest, to name but a few, are lined with cones, lava flows and ash deposits.

Besides its would-be tuyas, Valles Marineris displays a wide gamut of volcanic deposits, including cones and ash layers at the foot of the cliffs, vent-looking fissures, and ridges that look like feeder dikes. Fields of small pits might even be *pseudocraters* – steam blowouts caused by lava flows when they creep over water-soaked ground.

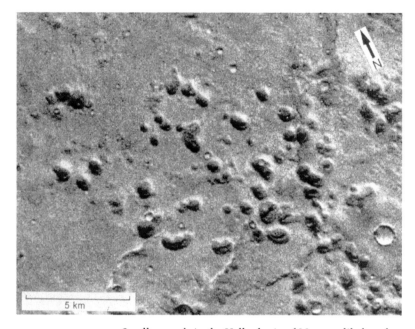

FIGURE 5.24 Small mounds in the Hellas basin of Mars are likely to be small volcanic cones and domes. Some show central pit craters, others elongated troughs that are reminiscent of breached cones on Earth. Deep basins like Hellas might be among the first provinces to be explored by manned missions. Credit: NASA/JPL, courtesy of the Lunar and Planetary Institute.

Pseudocraters are also found in other locations on Mars, notably in the young lava fields of Amazonis and Elysium Planitiae. These small low-profile cones, up to 250 m in diameter, pepper the landscape. They do not line up along faults, as do "real" volcanoes, but cluster in random groups, blistering the lava flows wherever they ran over buried lenses of ice or pools of ground water. Pseudocraters – also called "rootless cones" – are therefore reliable indicators of the presence of water near the surface. This is yet one more way in which volcanism sheds light on the environmental history of the red planet.

Present missions to Mars, starting with the 2004 exploration rovers, are targeting lowlands where ancient sedimentary deposits offer the best prospects of finding fossils and other biological clues. But the emphasis is bound to shift to young volcanic areas, like the

FIGURE 5.25 After the initial reconnaissance by automatic probes, manned expeditions will one day roam the volcanic landscapes of Mars. In this view, a team of astronauts has landed near a breached volcano and climbs the embankment of a lava flow. Credit: artwork by William K. Hartmann.

Elysium plains and the Tharsis shields, when it comes to understanding the recent past and present environmental conditions. When humans land on Mars and set up bases, these young volcanic areas will also attract attention. Not only will they interest geologists for their own sake but their higher-than-average heat flow might well have promoted groundwater circulation. Hydrothermal deposits might contain relics of Martian life, or at least clues about the underground chemistry. Salts and mineral ores could also be found in these volcanic environments.

For all these reasons, it is likely that future manned bases will develop at the foot of Martian volcanoes. It is tempting to imagine the first hikes and exploration campaigns that astronauts will undertake on these spectacular sites, and we use our poetic license to describe their visit to five prominent volcanoes of Mars in the following chapter.

6 A tour of Martian volcanoes

Olympus Mons

Longitude:	133°W
Latitude:	18°N
Dimensions:	840 km × 640 km
Volume:	2.4×10^6 km³
Relief:	21.9 km
Elevation:	21.1 km
Slope:	4–6°
Caldera diameter:	91 km × 72 km
Caldera depth:	3.2 km
Estimated age:	Upper Amazonian (100 million years)

FIGURE 6.1 Approaching Olympus Mons from the north. Credit: NASA/JPL-Caltech/MGS (MOLA), courtesy of Brown University.

Olympus Mons is the star of Martian volcanoes. It is by far the tallest edifice in the Solar System – towering 22 km above the plains – and the most voluminous: 2.4 million cubic kilometers.

The base of the shield shows steep cliffs and scarps. The flanks rise in stepwise fashion – a succession of sloping ramps and flatter terraces – up to a broad summit crowned by a complex 80-km-wide caldera. The overlapping flows testify to a long history of magma infilling, eruption and collapse. Lava flows decorate the flanks and many can be traced draping over the cliff and out into the surrounding plains, where they alternate with debris flows that attest to major landslides, known as the *aureole* deposits. Particularly well developed to the north, where they extend 1000 km into the plains, these gigantic lobes of jumbled terrain most likely resulted from the collapse of the lower flanks of the volcano, sliding off the edifice along

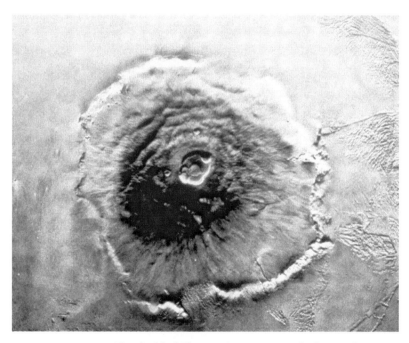

FIGURE 6.2 The shield of Olympus Mons. Notice the furrowed terrain – known as the aureole deposit – to the east (right). Credit : NASA/JPL-Caltech, courtesy of Brown University.

major faults and leaving fresh scarps and towering cliffs in their wake.

The scarps and jumbled features are reminiscent of submarine landslides around Hawaiian volcanoes. If the hypothesis of an ancient shallow sea on Mars is correct, the northwestern slope of Olympus Mons might have been submerged, explaining the magnitude and the orientation of the collapse. Another hypothesis is that the flow-like debris are glacial deposits. Climate computer models show that when the axial tilt of Mars reaches high values – as it periodically does – ice sublimes off the polar caps and some of it redeposits at low latitudes, along well-exposed major reliefs, such as the western base of Olympus Mons.

Olympus Mons was certainly active over a long period of time to reach its record size, and there is evidence that this activity persisted into the very recent past. Impact crater counts, performed on the entire shield, indicate that the outermost layers of lava coating the edifice average 100 million years or less, making Olympus Mons the youngest Martian volcano. In close-up, some areas are younger yet. The caldera displays so few impact scars that its age is more likely to be on the order of 30 million years. The greatest surprise comes from individual lava flows halfway down the flanks, imaged by the telephoto camera of *Mars Global Surveyor*. The flows look so fresh and undisturbed that they could not be much older than five or ten million years. Hence, Olympus Mons might still be active today, although its eruptions would now be so rare that we can only dream of catching one in the act. Finding evidence of recent activity on the volcano – such as hot spring deposits that might harbor Martian fossils – will nonetheless rank high among the objectives of future explorers.

Itinerary

Approaching Olympus Mons from the air will be a visual enchantment to future astronauts, especially if they fly in from the northwest. Skimming over the low northern plains – probably the floor of an ancient seabed – the pilots would spot the giant shield from afar,

FIGURE 6.3 Lava flows cascade over the basal cliff of Olympus Mons. Note the "railroad track" pattern of lava channels with levees. Credit: NASA/JPL-Caltech/MGS-MOC (Malin Space Science Systems).

warping the Martian horizon with its broad silhouette, 22 000 meters tall. Clouds of tiny ice crystals might hang in its wake, gently blowing in the late summer winds.

Less than 1000 km from Olympus Mons, the ground below the spacecraft would turn from flat plains to the gigantic lobe of crumpled terrain that spreads outward from the base of the volcano. The tongue of debris displays a series of ridges and troughs, like giant waves frozen in time, spaced 5 to 10 km apart. The comparison with underwater landslides is tempting, suggesting that the lower flanks of Olympus Mons literally collapsed – perhaps under the stress of a bulging magma chamber – and flowed out into the plains. It is difficult to imagine 100 000 cubic kilometers of lava and ash breaking loose from the base of the volcano and sliding down in a thunderous roar, displacing any ocean before it, like a mop pushing aside a puddle of water. The landslide probably reached velocities in excess of 200 meters per second, but even at such a speed, it would have taken close to an hour to spread down the slope and come to a rest.

As we near Olympus Mons, the scar of the giant landslide fills our windows, an escarpment 8000 meters tall – as high as the Himalayas – rimming the base of the volcano. We fly along the towering cliffs, until we reach a landing spot at the foot of the volcano,

FIGURE 6.4 Young lava flows on the slope of Olympus Mons. The absence of impact craters points to a very young age for these flows, perhaps less than 10 million years. Credit : NASA/JPL-Caltech/MGS-MOC (Malin Space Science Systems).

where lava has cascaded over the rim and built a sloping ramp that offers a viable access from the plains onto the shield. This hummocky ramp is the starting point of our roving expedition up the volcano in our pressurized "Marsmobile." We have barely begun our journey, heading up the slope, when we encounter a first example of the tremendous variety of volcanic landscapes on Olympus Mons.

Twisting channels 50 meters wide, framed by tall levees, meander down the slope and fan out into small deltas.We note the similarity with pyroclastic flows on Earth – the expression of explosive volcanism – and follow the channels up slope, until we reach two bowl-shaped craters, 500 meters wide, that appear to be the source of the fiery avalanches. Perhaps a column of magma encountered water-rich layers and set off the explosions.

We press on up the steep ramp until we reach the beginning of the shield proper, 8000 meters above the plains. In contrast to its steep pedestal, the shield's lower flanks slope up a gentle two to three degrees. Without satellite navigation, it would be difficult to decide which way is uphill, given that lava flows fan out across the shield and create their own local topography – swells 30 to 40 m high and 3 to 4 km wide, that mask the overall slope.

To hikers accustomed to the citrus groves, rice paddies or sugar cane fields of terrestrial volcanoes, Olympus Mons would look unbearingly desolate: not a shrub, not a blade of glass to break the monotony of a mineral world frozen in time. The only motion might be the sifting of red dust between the jumbled rocks, funneled down slope by the wind.

Many a lava flow displays a central channel, tens of meters wide, which funneled the molten rock down slope. In places, panels of solidified lava arch over the channel, remnants of what were once roofed-over lava tubes, before landslides and impacts shattered them to pieces.

Some lava fields are remarkably intact, with sharp boundaries and long sections of roofed-over tunnels. Large impact craters are conspicuously lacking, suggesting that these areas are extremely young.

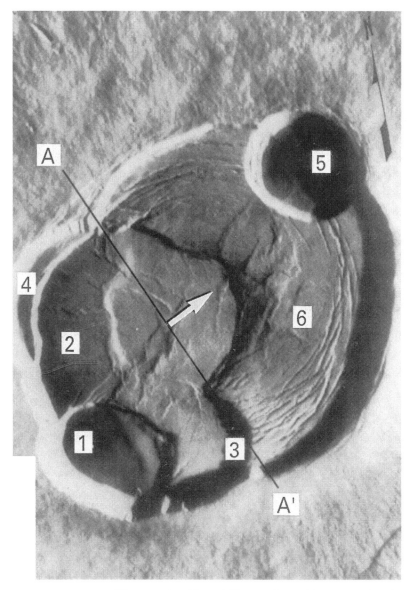

FIGURE 6.5 The summit caldera of Olympus Mons. Pits are numbered 1 to 6 from youngest to oldest. Credit: NASA/JPL-Caltech, courtesy of Dr. Thomas Watters, from 'Distribution of strain in the floor of the Olympus Mons caldera,' T. R. Watters and D. J. Chadwick, LPSC XXI, 1990.

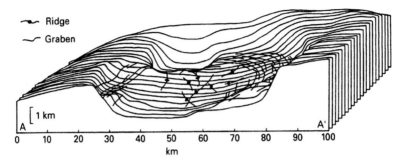

FIGURE 6.6 Topographic profile across the Olympus Mons caldera, along AA′ line, looking to the northeast. Extensional faults (graben) are bunched on the periphery of the caldera floor and compressional faults towards the center. Credit: see Fig. 6.5.

We would take special care in choosing and bagging samples from these flows for dating back in the laboratory.

Picking our way through lava channels and boulder fields, covering little more than 30 km a day, it would take us over a week to reach the summit. We wouldn't fail to notice half a dozen major changes in slope along the way – steeper ramps alternating with flatter terraces – that give Olympus Mons its characteristic bumpy profile. Arguably, the episodic bulging of the magma chamber buried deep inside the shield inflates and destabilizes the flanks, causing entire sections to slide down part of the slope and stack up as terraces.

As we near the summit, the slope steepens again. Summit eruptions discharge less magma than lower flank eruptions and build faster-cooling, stubbier flows that pile up around the vents.

Deep fissures begin to show up, breaking up the ground and hampering our progress as we approach the collapse zone of the central caldera. The swells of local lava flows mask the horizon, and the crater rim appears without any warning. Suddenly the ground drops 2000 meters below our feet – a sheer cliff curving left and right out of sight, with no sign of the opposite rim, so huge is the volcanic cauldron.

The caldera of Olympus Mons, averaging 80 km in diameter, is a complex collapse structure, caused by the withdrawal of magma

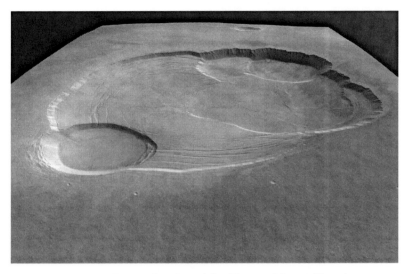

FIGURE 6.7 Perspective view of the Olympus Mons caldera rim, photographed by *Mars Express*. Credit: ESA/DLR/FU Berlin (G. Neukum).

from below the summit. The down-faulting occurred in several stages, affecting different areas at different times, as indicated by six overlapping craters with floors at different elevations. Apparently, the caldera first collapsed in one piece, before activity focused under its northeastern corner. There, further magma withdrawal led to the formation of an overlapping, deeper pit. Eventually, magma motion shifted to the western half of the caldera. There, a stepwise collapse ended with the creation of the southern, most recent pit, 3000 meters below the caldera rim.

After collapsing, caldera floors are often resurfaced by lava flows, when magma rises again, squeezing through the bounding fracture zones and flooding the crater. Olympus Mons is no exception: its caldera floor is coated with dark veneers of lava, buckled up in places by wrinkle ridges. We can imagine a time, perhaps not too distant in the past, when the giant caldera harbored lava lakes glowing in the Martian night.

On volcanoes on Earth, the lava level can occasionally rise to the rim of the caldera and spill out onto the volcano flanks, but this does not seem to have occurred on Olympus Mons. By flying around the rim – a Martian "hopper" would come in handy at this point – we would notice that the layering at the top of the cliffs shows a complex weaving of individual lava flows, rather than a continuous sheet of lava, as one would expect from a global spill-out. More realistically, lava lakes played at the bottom of the caldera, without rising too far up the walls, but they likely covered large areas to great depths. A scarp, slicing through the main floor, shows no layering down to a depth of at least 500 meters. The lava lake that pooled in the caldera was at least that thick.

The scarcity of impact craters on the caldera floor attests to its young age – perhaps less than 30 million years. But activity seems to have deserted the summit caldera today, and indeed we noted during our hike that the youngest lava flows are found halfway down the flanks, flush with the buried magma chamber. Arguably, there is still molten material deep inside Olympus Mons, but it will take infrared imagery, seismometers and tiltmeters to locate the pockets of magma and predict areas of future activity. Although new lava outbreaks are unlikely during our lifetimes, such areas will be surveyed for hot springs or any venting of gas that would open windows into the heart of Martian volcanism.

Arsia Mons

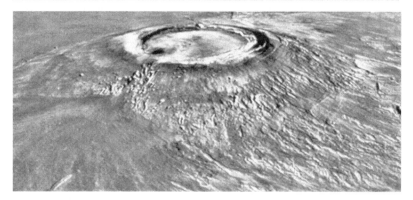

FIGURE 6.8 Perspective view of Arsia Mons (10:1 vertical exaggeration). Credit: NASA/MOLA Science Team, produced by the Goddard Space Flight Center.

Longitude:	121°W
Latitude:	10°S
Dimensions:	461 km × 326 km
Volume:	$9 \times 10^5 km^3$
Relief:	11.7 km
Elevation:	17.7 km
Slope:	1–5°
Caldera diameter:	138 km × 108 km
Caldera depth:	1.5 km
Estimated age:	Upper Amazonian (700 to 40 million years)

Arsia Mons is the southernmost of the great Tharsis volcanoes – a foursome that includes Olympus Mons and the NE–SW line-up of Ascraeus, Pavonis and Arsia Montes. Among Martian volcanoes, Arsia ranks third in terms of elevation (17.7 km) and in terms of volume (close to one million cubic kilometers). The shield appears to be relatively compact, because its lower flanks are surrounded and cut off by younger lava fields. However, some of Arsia's distal flows can be traced 2000 km from the summit.

The volcano is split down the middle by a prominent rift zone that follows the regional NE–SW trend. The rift segments are the source of abundant lava flows that fan out onto the lower flanks. Arsia is crowned by a giant, circular caldera. Over 110-km wide, the structure collapsed in one piece, contrary to most other calderas on Mars, which are made up of several overlapping pits. Arsia's caldera is bound by a set of semi-circular faults and terraces that drop down to a remarkably flat and smooth lava floor.

Crater counts on the shield and within the caldera yield a range of ages that underscore the volcano's complex and drawn-out history. Arsia was long believed to be the oldest of the four giant shields of Tharsis, with an average surface age of 700 million years. It appears today to have undergone a much younger cycle of activity that led to the eruption of the voluminous lava fans out of the rift zones, and to the contemporaneous collapse of the summit caldera. The last lava flows inside the crater might be as young as a few tens of millions of years and rank among the youngest volcanic units on Mars.

Itinerary

The exploration of Arsia Mons is a major undertaking, since important clues as to its complex history are scattered in the distal plains, on the shield's flanks, upper rift zone, and inside the summit caldera. A rocket "hopper" would be the best way to reach these far-flung locations and we will begin our geology tour with a visit to the distal lava plains, northwest of the volcano (at 2° S, 130° W). These smooth plains have intrigued geologists since the early 1990s, when the area was shown to absorb radar waves with uncanny efficiency and was for this reason nicknamed the "Stealth" region.

Once on the ground, the "Stealth" area turns out to be a gigantic field of sand dunes. Framed by lava flows, the dune field covers 1700 km² – the area of Death Valley – and packs in long waves of sand, up to 20 m in height, spaced roughly 500 m apart. A cover of fine ash sticks to the sand and holds it in place, preventing the grains from hopping downwind.

We would bag samples of the precious sand, in order to analyze and confirm its volcanic origin. Indeed, there are no flood channels in the vicinity, no water action that could have ground up the rocks into small grains of sand. Only an explosive eruption from a nearby volcano – namely Arsia Mons – could have shattered magma into sand-sized particles.

We return to our "hopper" and fly up the volcano, on the lookout for any vents or explosive craters that could explain the sand dunes. We see mostly lava flows in the plains below us. Many display central channels and look like railroad tracks in the grazing sunlight – spillways a couple of kilometers wide, framed by levees, which once carried molten lava down the slopes. We also see closely spaced ridges. These look like moraine piles left by a glacier that melted away. Ice glaciers on the flanks of equatorial volcanoes? Some climate models do suggest icy precipitation in the equatorial zone of Mars at certain periods. We would want to land in this mysterious terrain and look for traces of ice.

After this stop, we resume our flight up the volcano and soon run into the great rift zone that bisects Arsia's upper flanks. Towering cliffs line the spectacular canyon that we choose to follow towards the summit. As we near the top, we notice a 500-meter crater on the edge of the rift's headwall, looking very much like an explosion feature, and we come in for a landing on the narrow ledge between the crater and the plunging cliff face.

As we disembark for a closer look, the bowl-shaped crater reminds us very much of a maar – an explosion pit created by the interaction of magma and water. This would explain its location along the rift zone, which the magma followed to the surface, as well as the thick deposits of fall-back ash that mantle the crater and the surrounding landscape. From our position on the spur, we also have a good view of the cliff face, a natural cross-section of the volcano's flank. The top layer is a 50-meter-thick blanket of ash overlying the bedrock – presumably the product of the maar crater nearby, or of other explosive pits along the rift zone.

Leaving the cliff site, we fly the last few thousand meters up the

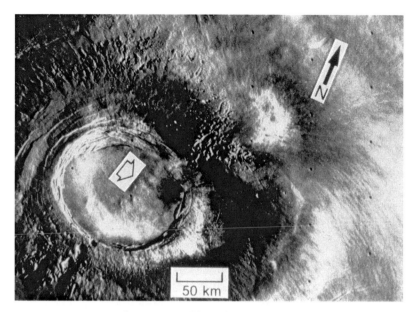

FIGURE 6.9 The summit caldera of Arsia Mons is 100 km wide. A fan of lava flows spreads out onto the northern flank. Notice the row of small domes on the caldera floor (arrow). Credit: NASA/JPL-Caltech, courtesy of Lunar and Planetary Institute.

volcano and over the top, to pick a landing spot inside the caldera. The huge summit depression – over 110 km in diameter – is arranged as a succession of ring scarps and terraces dropping down to a wide lava-covered floor, one thousand meters below the rim. The summit of the volcano apparently collapsed in one piece, as the underground magma chamber emptied out – feeding perhaps the large flank eruptions before caving in.

The collapse created a set of graben that constitute the caldera's outer rim. These ring faults are two kilometers wide and 75 meters deep with steep walls. In many places, we notice that lava erupted on their floors, piled up, and eventually spilled out to flood the surrounding terrain. We follow such a flow as it runs down a ten-degree slope, dropping 200 meters to the last terrace that overlooks the caldera. The lava spreads out across the terrace, but stops short of reaching the crater floor. Our hopper skims over the terrace and

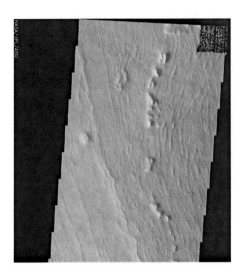

FIGURE 6.10 The western flank of Arsia Mons is covered with a thin, ridged deposit that could be moraines of volcanic ash dropped from a receding glacier. Credit: NASA/JPL/Arizona State University.

down the final slope, aiming for a smooth landing spot on the caldera floor.

Although Arsia's caldera is remarkably level, it is still rough on a small scale, with flow fronts several meters high and lava channels striking across the landscape like railroad tracks. Small impact craters pepper the surface, but we fail to detect any collision scar larger than a football field. In fact, we count ten to fifty times fewer craters than we would during a lunar landing, an indication that the Arsia lava fields are ten to fifty times younger than the mare flows on the Moon. They would range only 40 to 200 million years in age. This is remarkably young when one recalls that early studies tagged all Martian volcanoes as being *billions* of years old!

We land our hopper at the foot of a low hill. Its crest is split open by an eruptive fissure. As the dust settles, we embrace a view reminiscent of the lava plains of Arizona or New Mexico. Because erosion on Mars is one thousand times slower than on Earth, lava flows that are millions of years old appear to us as if they erupted only a few thousands of years ago. A first reconnaissance trek across the vast caldera will take weeks – it is the size of Yellowstone National Park. There are fissures to explore and domes to climb. Rocks to measure

FIGURE 6.11 Lava flows on the slope of Arsia Mons. Credit: NASA/JPL/Arizona State University.

and to date. How voluminous were the eruptions? Did Arsia build itself up in a dozen or so enormous eruptions, each time a "blob" of its mantle plume reached the surface? Thousands of years of violent activity, separated by millions of years of quiet?

Perhaps our expedition will tell.

Alba Patera

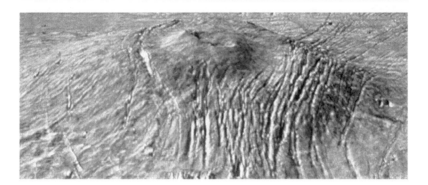

Longitude:	110°W
Latitude:	40°N
Dimensions:	1150 km × 1015 km
Volume:	1.8×10^6 km^3
Relief:	5.8 km
Elevation:	6.8 km
Slope:	0.5–1.5°
Caldera diameter:	138 km × 106 km
Caldera depth:	1.2 km
Estimated age:	Hesperian/Early Amazonian (1.5–3 billion years)

FIGURE 6.12 Perspective view of Alba Patera (11:1 vertical exaggeration). Credit: NASA/MOLA Science Team, produced by the Goddard Space Flight Center.

Alba Patera is the largest volcano on Mars, and for that matter the largest volcano in the Solar System, in terms of area and volume. Olympus Mons is taller but not as wide. The apron of lava of Alba Patera spans at least 1000 km in diameter, covering an area the size of France and Spain put together. Some estimates extend the lava shield to an overall span of 2700 km. The distal flanks of the volcano are so flat that its boundary is hard to trace.

Despite its great size, this is a "pancake" of a volcano: Alba Patera rises less than 6000 meters above the plains. Arguably, such

flatness is due to the massive outpourings of lava that spilled far and wide across the plains, and could not stack up to build a dome (except, as we shall see, in the central part of the volcano). The longest individual flows can be traced 900 km down the shallow slopes.

A prominent ring of fractures girdles Alba Patera halfway up the flanks. These large down-dropped graben were most likely created by the tearing apart of the edifice as it spread under its own weight. The fractures look like wrinkles around the corners of some gigantic, cyclopean eye. Past the deep troughs, the central part of the volcano – the "eyeball" itself – rises at a steeper gradient towards a broad, rounded summit, where a dark pupil stares out into space: a shallow depression, 130 km in diameter, containing a pair of subdued calderas. Sinuous lava flows are found in this area, trickling down the summit flanks. They represent the latest eruptions from this long-lived volcano.

To reach such gigantic proportions, Alba Patera must have been active for a considerable amount of time, roughly 3 to 1.5 billion years ago (throughout the period known as the Hesperian and into the early Amazonian). Perhaps some late stage eruptions occurred more recently in the summit area, where lava flows are much less disturbed by impact craters than farther down the slopes.

Itinerary

Surveying a volcano as large as Alba Patera is not an easy task. Crossing the shield on foot is out of the question: a straight traverse from the lower flanks to the summit would take a couple of months, at best. Even roving vehicles would be stopped in their tracks by the steep graben which surround the central shield, like the moats of a fortress. A thrilling alternative is to drift across the landscape in a balloon. This type of transport is under consideration for light automatic probes, and we use a sized-up version for the sake of our excursion. Our balloon would rise and travel by day and land by night as its gas cools and shrinks. By laying anchor at each interesting location, we can disembark and study the site.

FIGURE 6.13 Plan view of Alba Patera, showing the array of north-trending faults curling around the bulb-shaped shield. Notice the smaller volcano Ceraunius Tholus in the lower right corner, above the scale bar. Credit: *Viking* mosaic (NASA/JPL-Caltech, courtesy of Brown University).

Setting off from the plains of Acidalia, 1500 km west of Alba's summit, we would want to travel in the spring, to take advantage of the season's moderate westerly winds. A ground speed of 10 meters per second would take us 360 km during a 10-hour day and, counting the nightly stops along the way, we would reach the summit in four days' travel.

The first day of the journey might be rather uneventful, as we drift over the flat plains of Acidalia. By the late afternoon, however, with the sun low on the horizon, the grazing illumination would highlight the subtle topography of lava flows that define the outer apron of Alba Patera. Pock-marked by a sprinkle of impact craters, these flows could be as old as three billion years, but they still look fresh – a consequence of the very low erosion rates on Mars.

After a night's rest, as we resume our eastbound flight, our radar readings indicate that the land below is rising, 700 km from the volcano's center. The slope is less than one degree – a slope as gentle as the continental rise of our deep ocean floors. For every kilometer we travel, the terrain rises less than ten meters.

The comparison with our ocean margins is fitting, since Alba Patera is built on the boundary that separates the northern lowlands of

FIGURE 6.14 Faults and row of volcanic pits (below top crater) on the flanks of Alba Patera. Credit: NASA/JPL/Arizona State University.

Mars from the highlands to the south. By analogy with the Earth, it is tempting to also attribute this contrast on Mars to oceanic and continental provinces, and indeed there is some evidence that water or mud pooled in the lowlands of Mars early in its history. If this were the case, then Alba's northern and eastern flanks sloped down to the water line.

There are places on Alba, notably in the north, where the surface is gouged by winding valleys that do not resemble lava channels. They often cluster in networks of parallel branches, as if they were carved by running water – water running directly on the surface or flowing underground and sapping the land above. Either way, hard lava cannot be so easily eroded, so perhaps these areas are covered with ash, a more friable material and one that could soak up vast amounts of water from the atmosphere.

Another surprise awaits us at the close of our second day – 400 kilometers from the top craters – as the sun sets behind us and our balloon cools and lowers us to the ground. Great linear troughs tear up the landscape, running north–south out to the horizon. These graben are a couple of kilometers wide and hundreds of meters deep with sheer cliffs – a heart-breaking obstacle for any land expedition. This is the closest we'll ever get to a canal on Mars, except that these deep troughs carry no water! They were most likely formed by the fracturing and stretching of the Martian crust, as magma rose to the surface. Apparently, the magma stalled before reaching the surface and cooled in place – an intrusion known as a dike. We notice rows of circular pits on the graben floors, funnel-shaped depressions the size of football stadiums. They look like sinkholes, and the most probable explanation is that they mark the spots where gas seeped up from the cooling magma and broke to the surface.

Our fourth and final day of travel takes us to the top calderas of Alba Patera. We fly over a second set of troughs cutting across the lava fields – the ultimate moat girdling the summit of the volcano.

The summit is a broad cap of lava, 350 km in diameter, that rises an extra 1500 meters to a final altitude of 6800 meters above "sea level." This central shield is composed of a radial array of short,

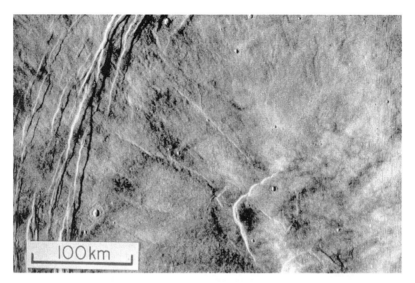

FIGURE 6.15 The summit of Alba Patera is girdled by an array of faults. The caldera (lower center of frame) is highlighted by a boomerang-shaped bright scarp. Credit: NASA/JPL-Caltech, courtesy of Brown University.

sinuous lava flows. In close-up, some areas have a lumpy, mottled appearance, like the moguls of a giant ski resort. Perhaps these are fields of wavy pahoehoe flows that were erupted at a slow rate and piled up in smooth sheets and lobes, unlike the more voluminous and turbulent aa flows that preferentially break up into a jumble of sharp boulders. The smooth subdued appearance of the lava field might also be due to a coating of ash erupted from the summit craters.

At the summit of Alba Patera lies a shallow depression roughly 130 km in diameter – larger than Yellowstone National Park. It comprises two calderas: a western one (95 km × 65 km) and a southern one (65 km × 55 km), separated by a low dome with a crater. The last eruptions of Alba were focused in this summit complex, creating a variety of landforms that come into view as we land our balloon in the southern caldera. Its walls are scalloped, attesting to multiple collapses of the magma chamber. Fine lineaments are visible along the escarpment: perhaps alternating layers of lava and ash. Wrinkle ridges, that were emplaced when the thick lava cooled and shrank, strike across the caldera floor, buckling the surface. Hollow impact craters and circular mounds – some look like cones of ash – cast subtle shadows across the landscape as the Sun sets over the western rim and the dusty sky turns a deep shade of lavender. At last we have reached the center of Alba Patera, the largest volcano in the Solar System.

Hecates Tholus

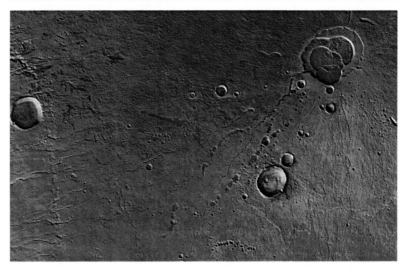

Longitude:	209° W
Latitude:	32° N
Dimensions:	187 × 177 km
Volume:	6.7 × 10⁴ km³
Relief:	6.6 km
Elevation:	4.8 km
Slope:	3–7°
Caldera diameter:	13 km
Caldera depth:	0.4 km
Estimated age:	Late Hesperian/Amazonian (1.7–0.5 billion years)

FIGURE 6.16
Hecates Tholus photographed by *Mars Express*. Credit: ESA/DLR/FU Berlin (G. Neukum).

Hecates Tholus is the northernmost volcano of the Elysium rise, which also hosts Elysium Mons and Albor Tholus. Averaging 180 km in diameter, Hecates Tholus is a relatively steep shield with slopes reaching 7°, flattening out near the top. A 12-km-wide complex caldera crowns the summit. Hecates is one of the most Earth-like volcanoes in terms of shape and size, comparable to the Big Island of Hawaii.

Hecates Tholus is named after the Greek geographer Hecateus. *Tholus* means hill in Latin, and refers to the steep class of Martian volcanoes, distinct from the flatter paterae and the mons shields.

The surface of Hecates Tholus shows little trace of lava flows. Instead, it is gouged by radial channels, which led early researchers to speculate that its eruptions were explosive in nature and generated ash flows that rolled down the flanks. It is now believed that the channels were carved by running water, but this still supports the notion that the flanks are made of soft ash. West of the summit, a wide area (75 km × 50 km) lacks surface detail, blanketed perhaps by an ash layer from a recent plinian eruption. Whereas the bulk of Hecates Tholus appears to be Late Hesperian in age (1.7 billion years), the terminal plinian eruption could be as young as a few hundred million years.

Itinerary

Hecates Tholus will no doubt be a top priority for volcanologists who explore the red planet. Not only is Hecates representative of the Elysium volcanic province – where igneous activity culminated halfway through the planet's history – but the volcano might also offer the best example of a plinian eruption on Mars, if the smooth area near the summit truly turns out to be a blanket of ash.

If they come in from the west, explorers will be met by a spectacular embayment that clips the shield's lower flank, as if an ogre had taken a bite out of the volcano. Over 40 km in width, the semicircular embayment is probably the scar of a giant landslide, possibly triggered by an impact. An open amphitheater, perched on the upslope part of the embayment, might be a leftover of the culprit crater.

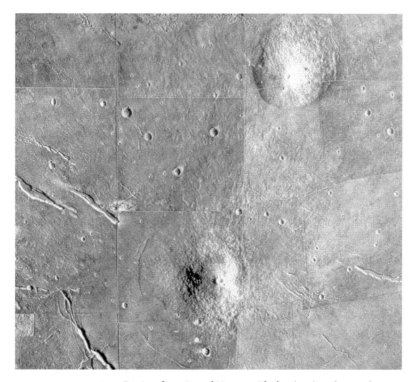

FIGURE 6.17 Regional setting of Hecates Tholus (top) at the northern end of the Elysium rise. Elysium Mons is at bottom. *Viking* mosaic. Credit: NASA/JPL-Caltech, courtesy of Brown University.

Smooth-looking lava fields lap up to the foot of the volcano and penetrate into the embayment. Riding up to the foot of the volcano with our rovers might be relatively easy. The embayment walls, over a kilometer high, would offer us a precious cross-section of the basal layers of the edifice. Indeed, a telephoto image taken by *Mars Global Surveyor* displays some fine layering at the top of the cliff, that might result from extensive ash falls. This is the sort of evidence we are seeking to confirm the explosive nature of the Hecates eruptions. A difficult climb up this steep section might reward us with an answer.

Channels emerge at the top of the cliff like hanging valleys. We follow one of these wide spillways uphill, looking for clues as to the nature of the carving fluid. Pyroclastic avalanches or mudflows? Hot

FIGURE 6.18 The base of Hecates Tholus presents steep cliffs overrun with lava flows. As we climb towards the "alcove" ridge (top of frame), we encounter fresh lava flows and small elongate vents. Credit: NASA/JPL/MOC/Malin Space Science Systems.

ash or cold water? Perhaps they were a combination of both: thick slurries of ash suspended in melt water, known on Earth as lahars or jökulhaups. These catastrophic outbursts occur on the slopes of tropical volcanoes after heavy rainstorms (as they did on Pinatubo in 1991) or in polar environments when eruptions melt the icecap of a volcano and flush ash-laden water down the slope.

On Hecates, a prominent gully leads us a distance of 40 km up the slope to the volcano's summit. Along the way of this long trek, we would want to check the embankments for any outcrops jutting out of the walls, in order to figure out if this relatively steep shield (3° to 7° slope) is built of layers of hardened ash, lava, or a combination of both.

The best proof of an ash eruption awaits us perhaps on the summit plateau, as we cross a ring of low ridges that marks the edge of an ancient filled-in caldera, 70 km in diameter. The western section is relatively free of impact craters, as if a recent layer of ash blanketed the underlying terrain. Identified on *Viking* images as early as the 1980s, the area is the size of Delaware (70 km × 50 km). If the ash rained down from a cloud, expelled from the central crater, this was a truly monumental eruption. We know that eruptive clouds are as tall as they are wide, based on the example of plinian eruptions on Earth. This one on Mars would have probably towered 50 km over the volcano. Such formidable heights are realistic in view of the low gravity of the Martian environment. If the rising cloud was propelled by water vapor, perhaps it was this water that created the channels down slope, as it rained out and penetrated the permeable ash.

The ash layer at the top of Hecates Tholus is a relatively recent event. Impact crater counts indicate an age younger than 300 million years, in contrast to the bulk of the volcano that averages closer to 2 billion years. If these numbers hold up, this would attest to the great longevity of Martian volcanoes.

Trekking across the ash blanket, we finally reach the complex central caldera that crowns the summit of Hecates Tholus. Four

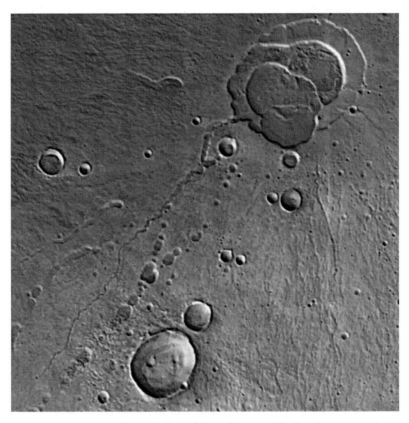

FIGURE 6.19 The complex caldera of Hecates Tholus shows several platforms due to multiple collapse of the lava surface. Credit: ESA/DLR/FU Berlin (G. Neukum).

craters overlap each other over a distance of 12 km: they become progressively smaller and deeper from north to south, as if the underlying magma – that withdrew its support from the summit – retreated southward and downward with time.

From our vantage point on the western rim, we stare down at a platform, nearly two kilometers wide, that marks the floor of the second crater in the series. The lava terrace stretches out to the horizon, where a ledge marks the steep drop to the floor of the next crater, out of sight.

The entire caldera complex is 400 m deep – comparable to the vertical drop across the nested pits of Kilauea crater in Hawaii. As we stare at the terrace of lava below, hummocky and dark in appearance, we can imagine a sequence of events similar to the 1790 eruption that shaped the Hawaiian caldera. During that historical event, an explosive outburst, probably fueled by groundwater, shot columns of ash out of the crater's ring faults, emptying the magma chamber and leading to a major collapse of the surface. Only later did a batch of degassed lava flood the crater floor.

On Hecates Tholus, we can imagine a similar scenario. Rising magma met the water table and turned into a billowing plume of fragmented ash that mantled the region west of the summit. Later, fiery fountains of red magma sprang up along the edge of the collapsed crater, flooding its floor with sheets of glowing lava. How many times did this cycle repeat? When did Hecates Tholus last erupt? These are some of the questions we will ask.

Hadriaca Patera

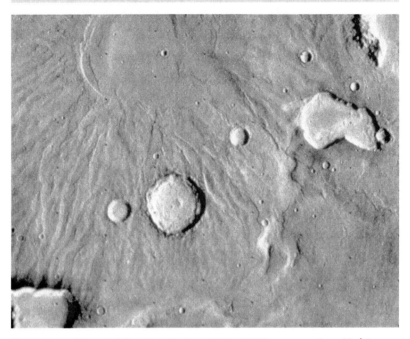

Longitude:	103°E
Latitude:	32°S
Dimensions:	550 km × 330 km
Volume:	1.6×10^4 km^3
Relief:	1.1 km
Elevation:	−0.6 km
Slope:	0.25–0.75°
Caldera diameter:	77 km
Caldera depth:	0.7 km
Estimated age:	Late Noachian/Early Hesperian (2–3 billion years)

FIGURE 6.20 Hadriaca Patera imaged by *Viking*. Circular caldera is at top left, with radial channels, impact craters and, at bottom, collapse depressions opening up into wide canyons (Dao Vallis). Credit: NASA/JPL-Caltech.

Hadriaca Patera is a type example of the eroded, subdued volcanoes of the southern hemisphere of Mars. It is perched on the rim of the Hellas impact basin. Its nickname is "Dandelion," in tribute to the numerous spokes that radiate from its central caldera like petals. These appear to be ancient water channels.

Hadriaca Patera is 330 km by 550 km in size, elongate in a north-east to southwest direction. The slopes of the edifice are so gentle, averaging half a degree, that the relief reaches little more than one kilometer over an area of 180000 km², the size of Ireland. The central caldera is 77 km across and its floor appears covered with ancient lava flows.

The spokes that radiate from Hadriaca's summit caldera turn out to be deep, broad valleys incised into the flanks of the volcano by liquid water in the distant path. The erosion runs so deep that many geologists believe the slopes are made of soft ash, rather than hard lava. Ash would point to the volcano's explosive behavior. Either the magma was inherently rich in volatiles or it encountered water-soaked ground on its way to the surface. Either way, the violent expansion of volatiles pulverised the magma into clouds of fine ash that rose high into the Martian sky. Some certainly collapsed to form pyroclastic flows that rolled down the volcano.

The eruptions of Hadriaca Patera took place early in Martian history. From the number of impact craters branded onto its flanks, the volcano is 2 to 3 billion years old, although the hidden base of the volcano might be older, and the late-stage lavas filling the caldera somewhat younger. The growth of Hadriaca thus spans a considerable amount of time, including a period when liquid water was present. Because heat and water coexisted, the volcano is a prime location where we should look for life. Fossils of primitive life forms could be hidden in the hydrothermal "plumbing system" of the volcano, or scattered in the ash beds.

FIGURE 6.21 Dao Vallis cuts into the southern slopes of Hadriaca Patera. The deep erosion points to soft layers of volcanic ash. Glaciers or running water carved the valley. *Mars Odyssey* Themis image. Credit: NASA/JPL/Arizona State University.

Itinerary

To explore Hadriaca Patera, one option is to land in Hellas basin and drive up Dao Vallis, a wide channel that cuts into the southern flank of the volcano.

Hadriaca is perched on the rim of Hellas basin for good reason. The giant impact that created Hellas so shattered the Martian crust that it created pathways for the deep magma to rise to the surface. Hadriaca likely straddles such a fault, but its apron of ash covers up any trace of it.

We land at the foot of the shield where Dao Vallis spills outs onto the basin floor. This vast outflow channel, 50 km wide at its mouth, tracks up the gentle slope of the volcano over a distance of 1000 km – the length of a river like the Rhine. A considerable amount of water was needed to carve out such a channel and it presumably sprang out of the volcano's flank – perhaps the discharge of its hydrothermal system or the melting of ground ice during pulses of heat.

As we proceed up the wide channel, we would not fail to check out the boulders that rolled down the flanks – samples of the putative, overhanging ash layers. Dark tracks streaking down the slope point to the best sampling locations. With a hand lens, we would perhaps find in these boulders flattened shards of volcanic glass, testifying to the high heat that fused the ash together, or else we might find droplets of clay, indicating that the ash came down tepid and wet from the eruption cloud.

On a greater scale, we would scan with field glasses or a telephoto lens the ash outcrops jutting out of the channel walls. Uniform, continuous layers run along the top ledge over tens of kilometers, which might point to extensive ash sheets (instead we would expect lava flows to show short overlapping lobes). Pasted on the valley walls, vertical tongues of sediment testify to some sort of veneer, deposited by wind or water.

Proof of flowing water is omnipresent. Not only was Dao Vallis carved by catastrophic outflow from the volcano in the distant past, but small gullies streak down the valley walls, which look much more recent. Could water still be seeping out of the ash layers? We would probably want to climb one of these gullies with our expedition rovers, a challenging but direct route that would hoist us onto the flank of Hadriaca Patera and launch us towards the summit crater.

With Dao Vallis behind us, the road to the top is easy to follow. Shallow broad channels point to the summit caldera, about 300 km distant. This would make it a ten-hour drive, at our top cruising speed of 30 km/h. Our pathway is choked with piles of ash that now conceal the channel floor, but we can guess that here Hadriaca's spillways were carved by groundwater, sapping the soft ash from below. We come across landslides – perhaps triggered by impacts nearby – that call for sampling stops and slow our progress. When we finally reach

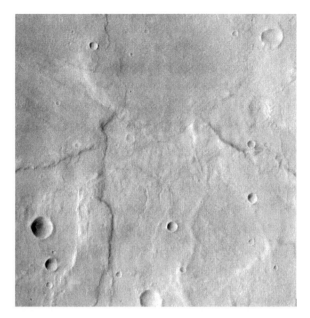

FIGURE 6.22 Lava spills from the edge of the smooth
caldera (top) onto the southern flanks (bottom). North is to
the left. Credit: NASA/JPL-Caltech, Arizona State
University.

the edge of the caldera, we are rewarded for our efforts: a steep scarp,
600 m high, towers over a vast arena of lava, that shows swelling and
buckling out to the horizon, as one might expect of a cooled and con-
tracted lava lake.

We could be standing on the edge of Kilauea crater in Hawaii,
although the scale is so much grander. Hadriaca's caldera is 77 km in
diameter and its far eastern edge lies out of view, but satellite imagery
shows that the caldera rim is lower in the east, and that the lava there
spills out directly onto the volcano's flank.

In order to drive down to the caldera floor, we need to circle the
rim until we reach the gently sloping overflow and use it as a ramp to
descend into the lava arena. But for the time being, spending the night
on the caldera rim is a fitting way to end the first day of a field trip on
Mars.

7 Volcanism on Venus

Among the planets, Venus is our nearest neighbor. It circles the Sun at an average distance of 108 million kilometers, a circular orbit that it covers in 225 days. Closer to the Sun and faster moving than the Earth, Venus laps us every 19 months. During the months when it is chasing us, Venus is visible as a bright evening star, before disappearing in the Sun's halo. After overtaking us, Venus emerges from the Sun's glare ahead of us, as an early morning star. Venus is also the world that resembles us most in terms of size. Its diameter reaches 12 104 km, only 652 km short of the Earth's 12 756 km. The area of Venus amounts to 90% that of the Earth's, and in terms of volume the ratio drops to 80%.

Venus is not the pleasant tropical paradise described by early science fiction writers, who were misled by its permanent veil of clouds and its "balmy" proximity to the Sun. Instead of a swampy haven for life, spacecraft showed Venus to be a hostile world, crushed by a hellish atmosphere 80 to 90 times the weight of the Earth's, and heated by a runaway greenhouse effect to a whopping 475 °C (750 K). It is so hot on Venus that its rocky surface should glow red in the dark.

AN ORIGINAL PLANET

Because of its permanent cloud cover, Venus long escaped the scrutiny of geologists, who were eager to find out if our sister planet experienced major volcanism and plate tectonics, comparable to Earth's. When radar surveys finally revealed the details of its surface, Venus turned out to be a volcanic wonder, replete with edifices of all shapes and sizes. Whole new types of volcanoes and igneous features were identified that baffled geologists and challenged our preconceived ideas of planetary evolution. This "menagerie" of volcanoes is

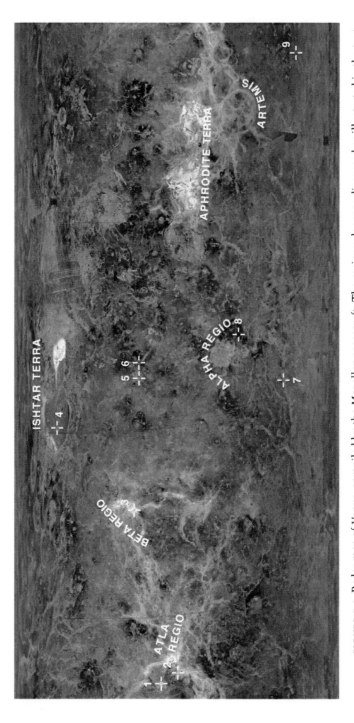

FIGURE 7.1 Radar map of Venus, compiled by the *Magellan* spacecraft. The major volcanoes discussed or illustrated in the text are indicated by numbers. (1) Sapas Mons; (2) Maat Mons; (3) Theia Mons; (4) Sacajawea Patera; (5) Sif Mons; (6) Gula Mons and Idem-Kuva Corona; (7) Ammavaru and Mylitta Fluctus; (8) Pancake domes; (9) Aidne Corona. Credit : NASA/JPL-Caltech.

all the more fascinating that it is beautifully preserved, due to the lack of rainfall and the limited rate of erosion on Venus.

Plate tectonics do not appear to operate on our sister planet. Instead, hot spot volcanism dominates the scene. Because of negligible erosion, the underground expression of these plumes of magma can be traced at the surface in the form of subtle swells and fault patterns. This record is not well preserved on Earth, which makes Venus particularly interesting.

On Venus, the pacing of eruptions also seems different to that on Earth. Volcanic activity appears episodic, rather than permanent. Great surges of volcanic hyperactivity seem to occur every few hundreds of millions of years, separated by long intervals of quiescence, as opposed to the constant creation and recycling of igneous crust on our home planet.

Last but not least, the high temperatures and pressures at the surface of Venus provide an original set of conditions that influence eruption styles and lava emplacement, in ways that enlighten our understanding of such processes in general.

For all these reasons, Venus beckons us to visit – an essential, albeit difficult destination for the serious volcano lover.

DESCENT INTO HELL

An opaque cloud cover shrouds the surface of Venus. Before the space age, the study of the planet was limited to its cloud deck and upper atmosphere. Spectral analysis of the clouds showed that these were composed of fine droplets of sulfuric acid, suspended in a dense atmosphere of carbon dioxide. What lay below was anyone's guess.

Preliminary measurements by *Mariner 5*, during the probe's quick flyby in 1967, showed atmospheric temperatures reach 430 °C (700 K), confirming early claims that Venus was prey to a monstrous greenhouse effect. In view of such hellish conditions, the first landing probes sent to Venus by the Soviets were built with the sturdiness of deep-diving submarines. Despite these precautions, their electronics were so battered by the hostile environment that *Venera 1* through *4* broke down during descent and ceased to emit before touchdown.

It was not until *Venera 7*, in 1970, that a probe finally survived to the surface. Landing on the night side of Venus, it radioed data for 22 minutes before its instruments failed. That was time enough to beam back the first weather report from another planet: the temperature reached 480 °C (750 K) and the pressure 9 MPa (90 atmospheres) – a crushing force similar to that experienced by a submarine under 900 m of water.

Two years after this initial success, *Venera 8* landed in the same region as its predecessor, but this time in plain daylight. During descent, its photometer indicated that the thick cloud cover ended 30 km from the ground. Carbon dioxide turned out to be the dominant gas (96%), followed by nitrogen (3.5%), with only traces of oxygen (0.01%), sulfur dioxide, water vapor and argon. Upon landing, the probe survived for close to an hour, made measurements of radioactive elements in the ground (uranium, thorium and potassium) and broadcast a new weather report: 470 °C (740 K), 9.3 MPa, and sluggish winds blowing at 1 m/s (3 km/h). The luminosity at the surface was so reduced by the cloud cover that the Soviets compared it to that of an overcast winter day in Moscow. Floodlights were added to the next generation of Russian probes, scheduled to take images of the surface.

Indeed, the first landings on Venus were conducted blind. There were no cameras. Only from the radioactive readings did the scientists guess that *Venera 8* had landed on some sort of alkaline lava, perhaps trachyte. But imagery was sorely needed and this was the challenge of the next pair of spacecraft.

THE BASALTS OF VENUS

Venera 9 and *10* reached Venus in October of 1975 and landed three days apart, at the foot of what would later be identified as a major volcanic plateau: Beta Regio. Each probe radioed back two black and white images of its landing site. *Venera 9*, perched on a 30° slope, revealed a landscape of flat rocks resting on a talus of cobbles and soil.

FIGURE 7.2 The ground of Venus, photographed on March 1, 1982, by *Venera 13* in the volcanic plains east of Phoebe Regio (7° S, 303° E). A cracked lava surface is visible in the foreground. Credit: USSR Academy of Science, courtesy of the Vernadsky Institute and Brown University.

FIGURE 7.3 Thin slabs of lava on the landing site of *Venera 14* (March 5, 1982). A spectrometric analysis by the Russian probe hints that the lava is tholeiitic basalt, similar to lava found on rift zones on Earth. Credit: USSR Academy of Science, courtesy of the Vernadsky Institute and Brown University.

It is now believed that the probe landed on the slope of one of the many faults that criss-cross the area (as later revealed by radar surveys). *Venera 10*, 1500 km to the south, imaged a landscape of layered rock "tiles," looking very much like the broken-up surface of a thin lava flow, with pockets of soil nested between the outcrops.

Gamma-ray analyses of the ground showed that the proportions of radioactive elements mimic those of terrestrial basalt. In particular, the moderate alkaline content (0.9% and 0.3% potassium on the *Venera 9* and *10* sites) mirrored the composition of basaltic tholeiites from the East African Rift.

In 1978, *Venera 11* and *12* failed to collect new images – a faulty circuit deactivated the instruments on both probes – but in 1982, *Venera 13* and *14* improved on their predecessors by broadcasting the first color pictures and detailed chemical analyses of the surface of Venus. Orange turned out to be the dominant color on the ground, due to the filtering of sunlight by sulfurous clouds and gas in the upper atmosphere. When global black and white imagery of Venus was later obtained by radar, it was colorized orange to reflect this finding. But geologists, eager to identify the lavas of Venus, worked hard instead to subtract the orange hue from the images, in order to ascertain the true color of the rocks. This correction showed them to be dark gray in essence, which is typical of basaltic unoxidized lava.

On the *Venera 13* site (Fig. 7.2 and Plate XIX), patches of gray lava stand separated by darker, granular material. In contrast, the smooth-looking rocks on the *Venera 14* site (Fig. 7.3) form a nearly continuous sheet with very little soil, perhaps because the lava flow is younger there. Radar imagery suggests that *Venera 14* landed on a field of youthful-looking flows, surrounding a 75-km-diameter volcanic shield.

Chemical compositions also differ slightly between sites, as measured by the X-ray fluorescence spectroscopy. On the *Venera 14* site, oxide proportions match those of tholeiite basalt with 49% silica and a mere 0.2% of potassium. In contrast, *Venera 13* lava shows less silica (43%) and more potassium (4%). On Earth, this would correspond to the fairly rare class of subalkaline basalt, as found on the slopes of Mount Vesuvius or the Eifel volcanoes of the Rhine graben.

Two more analyses broadened our view of Venus chemistry in 1985. *Vega 1* and *2* landed on the outskirts of Aphrodite Terra (180°W), nearly at the antipodes of the previous probes. Although they returned no pictures, they performed new analyses of the ground – one X-ray and two gamma ray spectroscopies – that pointed again to tholeiite basalt of hot spot, African rift "flavor."

To summarize, of the seven sites analyzed by the Soviet probes, five indicate chemical compositions typical of tholeiite basalt (*Venera 9, 10, 14, Vega 1* and *2*), one hints of subalkaline basalt (*Venera 13*), and one suggests an even more evolved subalkaline lava, similar to trachyte (*Venera 8*).

VOLCANIC GASES

From chemical analyses and imagery, then, it was obvious that the surface of Venus was essentially volcanic. But without imagery of the landscape, no one knew what the volcanoes looked like. Were they presently active, or on their way to extinction like the volcanoes of Mars? Scientists monitored what little they could see – the upper atmosphere above the cloud tops – for answers.

In the late seventies, astronomers had noted puzzling variations in the composition of Venus' upper atmosphere, revealed by spectral analysis. The amount of sulfur dioxide (SO_2) fluctuated slightly, and one explanation that came to mind was that a large eruption had taken place, replenishing the upper atmosphere in volcanic gases. This is a common occurrence on Earth, as we saw during the 1991 eruption of Mount Pinatubo in the Philippines, when an estimated 20 million tons of SO_2 were released. Sulfurous aerosols reached an altitude of 30 km and were traceable around the globe for an entire year.

In 1978, the American probe *Pioneer Venus* confirmed the variations of sulfur dioxide above the cloud tops. Its orbiting module observed a ten-fold decrease in SO_2 concentration from 1978 to 1983, along with a concurrent dissipation of the high altitude sulfurous

FIGURE 7.4 The atmosphere of Venus is composed mainly of transparent carbon dioxide. Sulfurous gases, attributed to volcanic eruptions, form clouds of sulfuric acid in the upper atmosphere of Venus. This art view of Maxwell Montes was created with *Magellan* radar imagery. Credit: David P. Anderson, Southern Methodist University, courtesy of LPI (Slide set: *It's a Dry heat: The Geology of Venus from Magellan*, compiled by Robert R. Herrick and Maribeth H. Price).

mist. One interpretation was that *Pioneer Venus* was monitoring the aftermath of an eruption: the fading of a recent sulfur "bloom" as it condensed and disappeared below the cloud tops.

The variations of SO_2 can be explained by a number of other processes. They could be chemical reactions in the upper atmosphere; mixing and overturning of gas layers under the influence of strong winds; or the formation and dissipation of clouds, masking and unveiling the underlying SO_2 gas. Volcanologists are also skeptical that eruptions would be strong enough to drive plumes of gas directly to such altitudes, given the crushing pressure of the atmosphere. The mystery still holds.

VENUS DROPS HER VEIL

Close-up pictures of lava and whiffs of sulfur were not enough. Volcanologists wanted to *see* the landscape and the putative volcanoes. The problem was cloud cover. The answer was radar.

Large radiotelescopes on Earth, namely Goldstone in California and Arecibo in Puerto Rico, bounced radar waves through the clouds and off the surface of Venus, generating a rough sketch of the relief along the equator. Two large swells, rising three kilometers above the plains, were named Alpha and Beta Regio and were suspected of harboring large volcanoes. Beta Regio, in particular, appeared to be flanked by large rift zones.

In 1978, the American orbiter *Pioneer Venus*, carrying a radar, crudely mapped out the topography of the planet and confirmed the existence of broad rises, rifts and plains. Two areas of high ground were identified: Ishtar Terra in the northern latitudes, the size of Australia; and Aphrodite Terra, a wide belt of ridges that wrapped a third of the way around the globe (15 000 km) and was half the area of Africa (see Fig. 7.1). As for Beta Regio, antipodal to Aphrodite, it was resolved into a dome-shaped uplift, capped by two volcano-looking shields, about 300 km in diameter and rising 4000 m above the plains. They were named Rhea and Theia Montes.

More detailed mapping by Earth-based radiotelescopes, in the years immediately following the *Pioneer* mission, showed the Beta shields to be dimpled by summit depressions and scoured by radial markings that ran down the flanks. Summit calderas and lava flows?

In 1983–84, two Soviet orbiters, *Venera 15* and *16*, took radar imagery one step further, piecing together a detailed mosaic of the northern latitudes of Venus. With a horizontal resolution of 1000 m and an altitude precision of 50 m, the probes revealed scores of volcanoes, which the Russian scientists classified in three size bins: large edifices (>100 km in diameter); intermediate volcanoes (20–100 km); and small ones (<20 km).

They also identified two large calderas in the high plateau of Ishtar Terra near the north pole – collapsed magma chambers that

FIGURE 7.5 Sacajawea Patera (64° N, 337° E) was one of the first features identified on Venus in early radar imagery. This high-resolution view by the *Magellan* probe shows Sacajawea to be an elongate caldera (225 × 150 km), rimmed by scarps and faults. Smooth lava flows show up dark, and rough lava flows show up bright in this radar image. Credit: NASA/JPL-Caltech (*Magellan*).

dwarf their counterparts on Earth. The one to the west, Colette, is 120 km wide. The one to the east, Sacajawea, boasts dimensions of 225 km × 150 km (see Fig. 7.5). Both are nested atop broad, shallow shields, arching 1000 m above the plateau baseline, and consist of a down-dropped central floor surrounded by ridges and troughs. Outside these bounding ring faults, bright radial markings run down the flanks over hundreds of kilometers – most likely rift zones and lava flows. When the *Magellan* spacecraft, in the 1990s, extended radar surveys to the entire planet, the list of calderas grew to 97 features, most of them falling in the 60–80 km size bin.

Venera 15 and 16 also imaged larger circular features that were unlike anything known on Earth. Named *coronae*, they consist of circular rings of deformed terrain averaging 200 to 300 km in diameter.

Contrary to a caldera, the central region of a corona is usually slightly higher than its surroundings. It contains a smattering of lava fields and channels, shields and domes. This central "lens" is often bounded by a raised rim and by a ring of fractures and ridges that point to intense tectonic deformation. More often than not, an outer trench or "moat" marks the boundary of the peculiar landform (see Fig. 7.10).

The two Russian probes identified some 30 coronae in the upper latitudes covered by their survey, clustered in belts along Ishtar Terra. Coronae were later found to be abundant near the equator as well, along the Aphrodite highlands, as well as in many other regions of Venus. Today, the corona family encompasses some 200 features, including a couple of gigantic landforms measuring 1000 km in diameter. One corona-like feature, Artemis Chasma, even spans 2100 km – half the size of Russia.

Venera 15 and *16* also discovered a special type of corona, so distinct as to deserve the separate name *arachnoid*, in view of the pattern of fractures that extends radially outwards, like a spider web. Numbering 265 planet-wide, arachnoids are typically smaller (<200 kilometers) and less outwardly volcanic than standard coronae, although they do seem to represent a variation on the same theme.

IN SEARCH OF PLATE TECTONICS

In summary, the first radar surveys of Venus sketched an unexpected portrait of our sister planet, with large low-relief volcanoes, giant calderas, and enigmatic circular features – arachnoids and coronae – surrounded by fractured terrain. Those who expected a world shaped by plate tectonics, similar to Earth, were hard pressed to find the familiar pattern of mid-ocean ridges and volcanic arcs that characterize the creation and destruction of crustal plates. But they did come up with several candidates.

Aphrodite Terra – the long chain of highlands girdling the equator – was tentatively compared to a mid-ocean ridge, where magma was presumed to rise along the axis of the massif. In its western half, the chain could be divided into putative rift segments, offset by faults. Profiles measured across the segments did show a rough symmetry about the central axis, with altitudes decreasing on either side (as is true with mid-ocean ridges on Earth). As for Beta Regio, geologists pointed out that its layout was very similar to that of the East African Rift on Earth, where continental crust is stretching, foundering and being replaced by oceanic crust.

If Venus did churn out new crust along the crests of Aphrodite Terra and Beta Regio, where did older crust sink back into the planet, as required by the mass balance of plate tectonics? The mountain belts around Ishtar Terra were suggestive of such a destructive plate margin. The southwestern belt in particular, Danu Montes, was bordered by a deep trough running 1000 m below reference level – the type of crustal down-flexing that is typical of a subduction zone on Earth.

The big picture, however, spoke against plate tectonics. The figures didn't match up. Firstly, there was no compelling evidence that the surface of Venus became systematically older as one moved outward from the putative rift zones. Secondly, mountain belts with

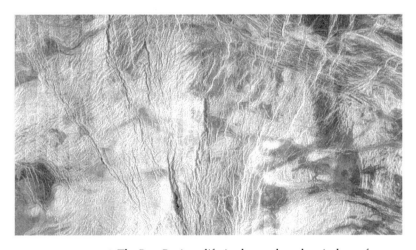

FIGURE 7.6 The Beta Regio uplift, in the northern hemisphere of Venus, is cut by a north–south rift, comparable to the Great African Rift on Earth. At the bottom of the image, the U-shaped downdropped valley is over 100 km wide. Credit: NASA/JPL-Caltech (*Magellan*).

trench-like margins did not add up to more than 3000 or 4000 km of potential subduction zones, versus 10000 or 20000 km of postulated spreading ridges in Aphrodite Terra. This amounted to a three to seven-fold discrepancy between estimated crust production and crust destruction. Hence, despite the fact that it was conceptually attractive, the Earth-like model of plate tectonics does not appear to fit Venus. Another mechanism is at play to evacuate the planet's internal heat, and the obvious alternative is hot spot volcanism.

On Earth, rising plumes of hot rock, turning to magma as they reach the surface, account for 10% of our planet's heat release, versus 70% for plate tectonics (and 20% for simple heat conduction). Without the efficient contribution of plate tectonics, Venus had to rely much more on hot spot plumes. Large volcanoes, coronae and arachnoids were perhaps the "blooming" heads of such plumes reaching the planet's surface. There still remained the mystery of where the older crust might sink on Venus, in order to achieve the balance of mass. Finding a suitable answer to this thermal and structural puzzle would be one of the major goals and achievements of the *Magellan* mission.

MAGELLAN EXPLORES VENUS

Launched by the Space Shuttle in May of 1989, the *Magellan* spacecraft achieved Venus orbit on August 10, 1990, and began its radar survey one month later. The images returned by the American probe boasted a horizontal resolution of under 300 m – a ten-fold improvement over the *Venera 16* and *17* imagery. Large volcanoes came into sharper focus and small volcanoes popped out of the background in droves.

In the northern latitudes, the *Venera* spacecraft had already identified over 20000 small shields and cones, two to three kilometers across. Planet-wide, *Magellan* brought the count to hundreds of thousands of features, complete with lava flows and channels. Many were clustered in groups of several hundred, reminiscent of volcano fields in Arizona, Idaho or Mexico, not to mention the innumerable volcano fields of our ocean floors.

FIGURE 7.7 Radar image of a medium-sized volcano, 90 km wide, located at 13.5° S and 315.5° E. Several lava fields overlap, with different textures and thus different radar brightness. The rough areas are bright. Credit: NASA/JPL-Caltech, courtesy of Brown University.

A planet-wide tally has come up with 647 such volcano fields on Venus, and there are many more areas where small volcanoes are loosely scattered throughout the plains. In a typical field, which is 100 to 150 km in diameter, there are 100 to 200 volcanoes spaced three to four kilometers apart. Most fields are found in the plains and basins of Venus, but some are found on the flanks of large volcanoes (like the "parasitic cones" of terrestrial shields), within rift zones and inside the fault rings of coronae. All in all, the total number of small volcanoes on Venus is likely to run in the millions.

Small volcanoes on Venus are commonly shield-shaped, although some are flat-topped like the seamounts of our ocean floors. They are often crowned by a summit pit crater, have shallow slopes and are surrounded by aprons of lava or ash, ranging from bright, radar-rough terrain to dark smooth patches.

The clustering of volcanoes into fields 100 to 200 km wide tells us something about the supply of magma to the surface of Venus. One hundred kilometers happens to be the estimated width of hot plumes on Earth and arguably on Venus as well. As they rise through the mantle and into the crust, such plumes experience melting, due to the drop in pressure. They release blobs of magma that trickle through faults and fractures to the surface, building the small cones and domes. Shield fields could therefore mark the emplacement and size of rising plumes. Models also tell us that the rate of magma production in such plumes is less than half a million cubic meters per year. At higher rates, the magma is expected to collect in large underground pockets – magma reservoirs – and yield instead large central volcanoes.

LARGE VOLCANOES ON VENUS

There are 168 volcanoes on Venus larger than 100 km in diameter. Most are found along rift zones and fracture belts, in regions of tectonic extension. Such regions are often topographic rises that reflect the uplift and thinning of the lithosphere over buoyant hot plumes. Large volcanoes preferentially develop at the intersection of several fault planes, as they do on Earth, which results in chimney-like feeding pipes.

Theia Mons, for example, is located at the intersection of three rift zones, criss-crossing the Beta Regio uplift (see Fig. 8.1). As for Ozza Mons in Atla Regio, it lies at the intersection of no less than five rift zones.

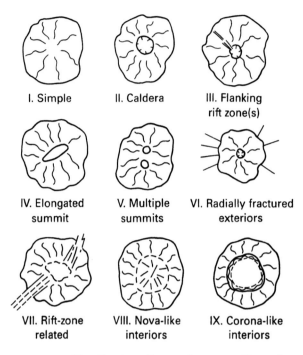

FIGURE 7.8 Classification of large volcanoes on Venus, featuring nine different types from simple shields (Type I) to shields with complex summits (Types II to V) and shields with major tectonic deformation (Types VI to IX). Credit: from J. W. Head, L. S. Crumpler and J. C. Aubele, 1992, Large shield volcanoes on Venus: distribution and classification. *Lunar Planet. Sci.*, XXIII, 513–14.

Of these 168 major volcanoes, half are over 400 km in diameter, and seven features exceed 700 km (Theodora Patera and Ozza Mons reach record diameters of 1000 km). But despite their great size, they are extremely flat. Their average relief is only 1500 m and only one is truly tall: Maat Mons, which towers 8000 m above the reference "zero" level.

On the other hand, Venus volcanoes are quite diverse in plan view. Nine different classes are recognized on the basis of summit geometry and tectonic layout (see Fig. 7.8).

The first two classes encompass all simple volcanoes with radial flows on their flanks and no caldera (Type I) or a simple caldera

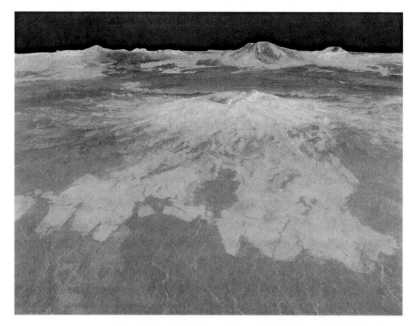

FIGURE 7.9 Sapas Mons is a large volcano located near the equator (8° N, 188° E) on the western edge of Atla Regio. The edifice is 400 km wide, but only 1500 m tall (10 × vertical exaggeration in this perspective view, based on *Magellan* radar imagery). On the horizon stands Maat Mons, the tallest volcano on Venus (8000 m). Credit: NASA/JPL-Caltech, JPL Multimission Image Processing Laboratory.

(Type II). A third category (Type III) comprises volcanoes with both a caldera and a prominent rift zone, as portrayed by Sif Mons. Other volcanoes have complex summits: elongated crests (Type IV), symbolized by Gula Mons, or even multiple peaks (Type V), the showcase example being Sapas Mons with its double summit (see Fig. 7.9). Two classes regroup volcanoes that are heavily marked by regional tectonics: those with radial fractures extending away from the shield (Type VI) and those straddling major regional rift zones (Type VII), exemplified by Beta Regio's Theia Mons (see Fig. 8.1).

The last two types comprise features that display central arrays of faults: radial arrays (Type VIII) and concentric arrays (Type IX). They are transitional classes, grading to original features that are found on no other planet: stellate fracture centers, arachnoids and coronae.

SPIDERS AND CROWNS

Annular features hundreds of kilometers in diameter abound on the surface of Venus. There is no doubt that they are igneous features. Many of them are studded with shields and domes and exude lava flows from their pervasive fractures. But their tectonic fabric and topographic expression indicate that the emplacement of magma occurs mostly underground in magma chambers, sills and dikes.

It is well known that the vast majority of magma that rises through a planet's crust actually "congeals" underground before reaching the surface. This is the case on Earth, where ten times more magma is emplaced subterraneously (plutonic rocks) than gets to

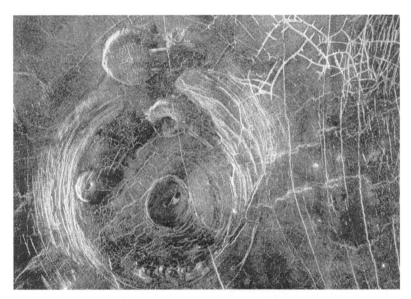

FIGURE 7.10 Aidne corona is a 200-km-wide circular structure in the plains south of Aphrodite Terra (59°S, 164°E). The corona is characterized by a ring of bright ridges and grooves, associated with volcanic lava flows and domes. Note the large dome to the north. Radar image by the *Magellan* spacecraft. Credit: NASA/JPL-Caltech, courtesy of Brown University.

erupt (volcanic rocks). When such batches of magma rise into the Earth's crust and deform it, radial and concentric fractures should develop on a large scale. Erosion, however, quickly erases the surface expression of these large intrusions. The lithosphere on Earth is also thicker than on Venus and mathematical models show that the tectonic fracturing due to a rising plume will be less obvious on our planet. Hence, the mechanism at work on Venus is of great interest to geologists.

These original, fractured landforms comprise some 208 coronae, 265 arachnoids and 160 stellate fracture centers, which all seem to represent variations on the same theme: that of underground plumes uplifting and fracturing the lithosphere.

Stellate fracture centers – once called *novae* – straddle topographic rises and are nested within arachnoids and coronae. They are characterized by a dense array of grooves that radiate from the center in a star-like pattern, hence their name. Their equivalent on Earth could be "radial dike swarms," such as the great McKenzie complex in Canada, where dikes extend 2600 km from the center. Stellate fracture centers might represent the initial stage of deformation caused by the impingement of a hot plume under the lithosphere.

Arachnoids are a variation on the same theme. They have stranger lineament patterns that resemble spider webs, including an annular center – similar to a corona – and radial ridges that extend outward and often merge with regional fracture belts. Arachnoids likely result from hot plumes impinging on previously fractured regions.

As for coronae, they are believed to be the full-blown expression of a plume that not only impinges on the lithosphere but also spreads beneath it, then cools and subsides. The lithosphere would first stretch to form a dome-like uplift, cracked by radial faults (stellate fracture phase). Then, the surface would undergo compressional, concentric deformation, as the plume spreads underground and its stress field moves radially outward.

Coronae show a variety of morphologies that appear to illustrate

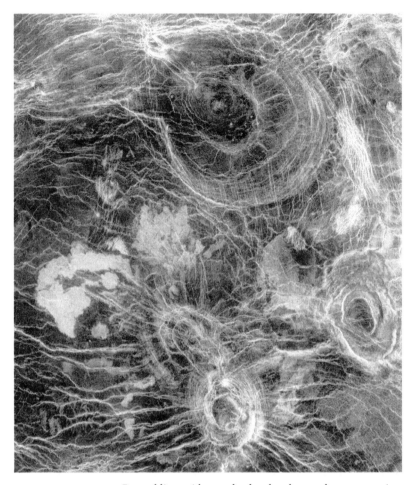

FIGURE 7.11 Resembling spiders and cobwebs, these volcano–tectonic features on Venus are named arachnoids. This field is centered at 40° N, 18° E, and contains several features ranging from 50 to 230 km in diameter. According to one theory, arachnoids are precursors of coronae. Credit: NASA/JPL-Caltech.

different phases of this subterranean spreading pattern. Early phases are dome-like features with little concentric deformation. More mature examples are circular to ovoid plateaus, with the outer slopes beginning to show annular fracturing. In the most evolved cases, the central plateau has subsided in a dimple-like basin, with a relatively elevated rim that shows concentric fracturing. A trench or "moat" often marks the outer boundary of the structure.

Physical modeling of this process yields time scales on the order of 200 million years for the blooming (stellate pattern), expansion (plateau corona) and relaxation (basin-and-rim corona) of a plume-driven structure on Venus.

Be it as it may, coronae are among the most original and intriguing features discovered on any planet. Besides their underground plutonic expression, their fractures have in many cases spilled magma out onto the surface in the form of vast lava fields, shield volcanoes and steep-sided domes. A landmark corona, Idem-Kuva (225 km in diameter) is thus featured in our selection of field trips (see pages 255–260).

PANCAKES AND TICKS

This family portrait of Venus volcanoes would not be complete without a description of medium-sized volcanoes (20–100 km in diameter). In this size bin, Venus also graces us with original landforms found nowhere else in the Solar System.

There are 289 medium-sized volcanoes on Venus. Some are simple shields with lava flows of uniform length adorning their slopes, like petals on a flower. Others are steep-sided domes, which count among the most intriguing volcanoes of our sister planet. Also known as *pancake domes*, these features are circular lenses of igneous "dough" with a puffed-up fractured crust (see Fig. 7.12). They resemble siliceous domes on Earth, which are built of viscous dacite or rhyolite lava that oozes out of the ground in one great batch. Lassen Peak in California is such an *endogenous* dome, as are the domes and needles inside the calderas of Mount Saint Helens, Martinique's Mount Pelée and other subduction volcanoes. Domes of hot oozing rock are often unstable during their growth, and experience glowing avalanches and the collapse of entire flank sections.

Avalanche scarps, hummocky terrain and tongues of debris likewise characterize the steep domes of Venus. Such "modified"

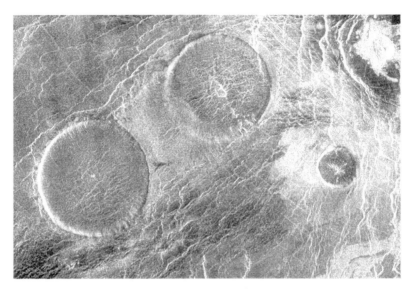

FIGURE 7.12 Two large (50-km-wide) steep-sided domes on Venus. Such pancake domes might have been formed by viscous, gas-rich magma stretching a brittle envelope. Note a smaller dome to the right and a bright impact crater in the upper right corner. Credit: NASA/JPL-Caltech, courtesy of Brown University.

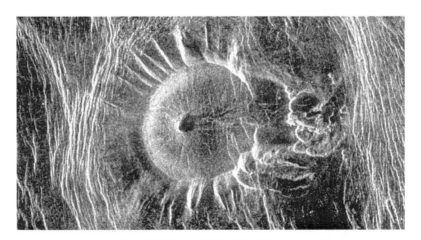

FIGURE 7.13 Eroded domes with leg-like ridges protruding on the flanks are nicknamed ticks in view of their insect-like appearance. Notice the dark pit in the middle of the flat summit. Credit: NASA/JPL-Caltech, courtesy of Brown University.

domes often have a fluted aspect with scalloped flanks – a succession of alcoves separated by ridges of more resistant material. Often, the central summit is slightly concave, as if eruptions had emptied out the magma chamber and caused it to sag. Such volcanoes are nicknamed "ticks" in view of their peculiar buggy aspect (see Fig. 7.13). Although Venus domes resemble siliceous domes on Earth, it does not necessarily follow that they are also siliceous. For one, their sheer volume – ten times that of their terrestrial counterparts – speaks against them being the last dregs of a cooling crystallizing magma chamber. Other processes on Venus could generate viscous dome-building material. Abundant gases might puff up the magma, even if it were basaltic, into a gigantic lens of "foam" that spreads radially out of the vent. The high pressure of the Venus atmosphere would prevent the foam from exploding into an eruption plume and confine it instead to a ground-hugging hemispheric flow. In order to settle the issue, one would surely want to explore and sample a pancake dome on Venus, a field trip that we also lay out in the next chapter (see pages 260–264).

LAVA FLOWS ON VENUS

We have focused so far on volcanic constructs: the shields, domes and coronae of Venus. But as is the case on all planets, the prominent form of volcanism on Venus is lava plains. Covering the vast lowlands and also perched amidst the highlands, lava plains might well account for 95% of the planet's volcanic output.

Many lava plains are formed locally. They resemble the landscapes of plain volcanism on Earth. Like the Snake River Plain of Idaho, they are peppered with small shields and channels and probably result from shallow melting directly below the surface. Elsewhere, large flows pour out of fracture belts, including those of coronae and large volcanoes, and spread out in fields hundreds of kilometers wide.

One example is Mylitta Fluctus (see Fig. 8.21). The flows originate from a broad shield straddling a fracture zone in the southern hemisphere, run down slopes as gentle as 0.1° and fan out at their extremities into deltas tens of kilometers wide (see field trip, pages 265–270). Such behavior implies a very low viscosity, comparable to that of lunar basalt and terrestrial *komatiite*. Fluid, metal-rich komatiite and tholeiite basalt are precisely the type of magmas that are expected to form from the partial melting of the Venus mantle, and they are compatible with most chemical analyses conducted by the *Venera* and *Vega* probes.

Some lava flows on Venus reach amazing lengths. Hildr channel holds the record, winding some 6800 km – the length of the Nile – from the highlands of Atla Regio to the plains of Atalanta. On Earth, even komatiite flows do not exceed 200 km, which bring geologists to propose even more exotic lava types on Venus. Carbonatite, for example, would stay molten over very long periods and distances at the high temperature that prevails on Venus, but chemists question the stability of carbon-rich magma at the surface.

The unusual length of lava flows is probably due to a conjunction of several factors, not only chemical but physical as well. On Venus, high temperature and pressure come into play. However, some of their effects are counter-intuitive with respect to our experience on Earth. One would indeed expect the heavy atmosphere, heated at 750 K, to act as a warming blanket, slow down the cooling rate of lava and allow it to flow over larger distances. But at the high temperatures at which basalt and komatiite erupt (1500 K to 1700 K), the 750 K surface temperature is not "felt" to be that hot and will not substantially slow down the cooling rate of the exposed material. On the contrary, the Venus environment might even speed up the cooling process. Because of its great density, the atmosphere has a high capacity of heat storage and transport, drawing large amounts of heat out of the lava and convecting it away. Calculations show that the thick atmosphere will cool exposed lava one and a half times faster than our thinner (albeit cooler) atmosphere does on Earth.

There are more surprises. By efficiently chilling the surface of the lava, the Venus atmosphere might nonetheless cause it to flow farther. Indeed, if an insulating crust forms rapidly over the lava

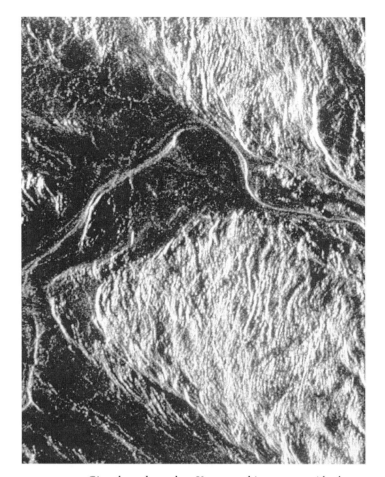

FIGURE 7.14 Giant lava channel on Venus, snaking across a ridged plateau north of Freyja Montes. The 2-km-wide channel takes its source in the mountainous cleft to the right and plunges 3 km to the dark plains. The nature of the fluid lava is unknown. NASA/JPL-Caltech, courtesy of Brown University.

stream, it will seal it off from the atmosphere and interrupt the cooling process. Hot magma would then flow unimpeded through roofed-over channels and lava tubes.

All in all, the high pressure at the surface of Venus appears to have a greater influence on the cooling behavior of lava than the high temperature does. On the other hand, temperature might exert greater control over the behavior of magma underground. Because the temperature inside Venus increases downward by an estimated 10 to 20 K per kilometer, it might reach 1000 K at 12 km depth and 1350 K at the base of the crust (30 km). Hence, magma rising through fractures will lose little heat to the encasing wall rock during ascent, much less so than in the Earth's colder crust, and thus collect in larger quantities. One would then expect larger magma reservoirs and dikes on Venus, a prediction that is borne out by the larger size of volcanoes, calderas and coronae on our sister planet.

The easy circulation and storage of magma underground might also explain why volcanoes on Venus have such low relief. According to hydrostatics, two factors should influence the altitude reached by rising magma. One is the density difference between the ascending melt and the encasing wall rock: the greater the density difference, the greater the ascent rate. The other factor is the depth of the magma source: the deeper the source, the higher the altitude it can ultimately reach. On Venus, both factors oppose edifice growth. Magma stalls in reservoirs close to the surface, where its density difference with the wall rock is minimal. The shallow depth of these reservoirs then prevents the magma from reaching substantial heights when it finally rises to the surface. It does not help either that lava flows over long distances, spreading edifices outward rather than allowing them to build up vertically.

ERUPTIONS ON VENUS

The collection of magma underground, nested in warm bedrock, and its rapid cooling by the dense atmosphere when it reaches the surface are two major effects of the peculiar Venus environment. The eruption process itself is highly dependent on atmospheric factors and we can question which of the typical hawaiian, strombolian, plinian or pelean styles of activity can occur on Venus.

The high-pressure environment has a profound influence on the behavior of volatiles inside the magma. It is the volatile content and its expansion into gas bubbles that drive explosive eruptions when magma nears the surface. On Earth, it takes less than 0.5% by weight of water vapor in the melt for bubbles to grow and blow magma apart at the surface and spray it out as lava fountains or columns of ash. At the much higher pressures that reign at the surface of Venus, comparable to those reigning on our ocean floors at depths of 800 to 1000 meters, it takes significantly more volatiles – several percent by weight – to attain similar results. Since the pressure is strongly altitude-dependent, the location of the volcano plays a role. In the lowlands, where the pressure is highest (10 MPa), 5% of carbon dioxide is needed to disrupt the magma, a figure hard to reach. In the highlands, where the pressure is lowest (less than 8 MPa), 2% of carbon dioxide is sufficient to trigger explosive behavior and such figures can theoretically be achieved, although opportunities will be rare and limited to the summit of the tallest volcanoes.

Plinian eruptions in particular, requiring the largest gas concentrations, would not only be rare, but also severely muted in

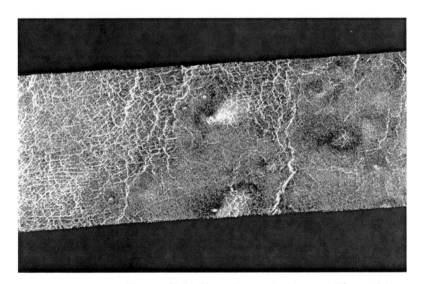

FIGURE 7.15 Two small shields with summit pit craters (1 km wide) in the plains of Guinevere (north is to the left). The dark pearly material covering the web-like reticulate plains is thought to be ash deposits. The bright wedge that extends right of the summits is probably bedrock, with the ash removed by the wind. Credit: NASA/JPL-Caltech.

character. The thick atmosphere dampens both ballistic flights and convection clouds. Blocks of lava shooting out of craters will decelerate forcefully and fall back close to the vent, and convecting plumes of ash will rise to only half the height that they would on Earth, other conditions being equal.

On the lower end of the scale, hawaiian eruption – droplets of magma propelled tens to hundreds of meters above a vent – can be achieved on Venus for reasonable amounts of volatiles (less than 1% by weight) and should therefore be relatively frequent. The size of fragments carried by these fountains might be larger than on Earth, because the greater density of the propelling gas has the capacity to lift heavier loads. This also applies to vulcanian eruptions, which occur when gas pressure builds up inside a plugged-up conduit. When the plug fails, the overpressure is released explosively, thrusting large blocks out of the vent. Here again, heavier blocks will be lofted than on Earth: "lava bombs" on Venus could reach diameters of 10 to 20 m (the size of trucks) and weigh thousands of tons.

Strombolian eruptions should also be fairly common on Venus. This type of eruption occurs when magma squeezes up tight chimneys at a slow rate, so that its gas bubbles rise faster than the magma and coalesce into clusters as they sweep through the magma column. Instead of a stream of small bubbles breaking at the surface without enough force to disturb the magma, one gets intermittent packages of coalesced bubbles, which puff up into large volumes when they reach the top of the chimney and blast blobs of magma above the vent. In this strombolian mode, there is no need for high concentrations of volatiles at the onset: the process concentrates whatever is available to periodically reach the disruption threshold. Strombolian activity is therefore a style of behavior well suited to Venus' volcanoes.

SPATIAL DISTRIBUTION OF VOLCANOES

Most of the Venusian volcanoes are now located and cataloged. We can search for patterns in their distribution, which might tell us something about the internal behavior of the planet. We have seen that there is no planet-wide grid of rift zones and crustal plates on Venus, as there is on Earth. The principal mode of volcanism on Venus is not plate tectonics, but hot spot activity, in which individual plumes pipe hot magma from the deep mantle to the surface. The distribution of hot spots on Venus is non-random, as a tally of volcanic centers clearly shows.

Firstly, there are more volcanic centers in the equatorial zones than near the poles, a pattern that shows up on Earth and Mars as well. Theoretical models show this to be a logical result of convection and heat evacuation inside a rotating sphere.

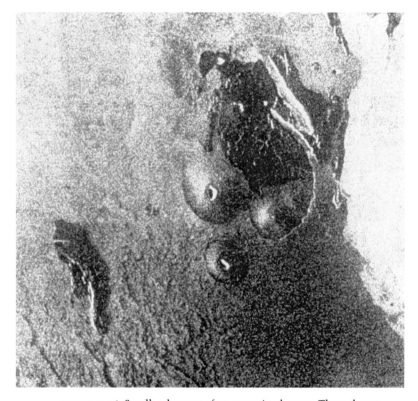

FIGURE 7.16 Small volcanoes often occur in clusters. These domes decorate the flank of Maat Mons in Ovda Regio (3°N, 195°E). The central edifice is 10 km in diameter and its height is estimated to reach 700 m. A radar-bright rough lava field lies to the east (right of image). Credit: NASA/JPL-Caltech, courtesy of Brown University.

There is also an obvious grouping of volcanoes and coronae in one broad region of Venus, where figures triple from 3 to 10 volcanic centers per million square kilometers. This concentration occurs over an area 13 000 km in diameter, located between longitudes 180° E and 340° E and latitudes 40° N and 50° S. Its nickname is the "BAT" zone, since it includes the regions of Atla, Beta and Themis. A second, less obvious grouping of volcanoes occurs in the opposite hemisphere between 0° E and 50° E from 0° N to 40° N. This second zone includes the regions Eistla and Bell.

This concentration of hot spots in one hemisphere is reminiscent of the patterns both on Earth and Mars. On Earth, hot spot activity is concentrated in the hemisphere that comprises the Atlantic Ocean, Africa and the Indian Ocean. On Mars it is the Tharsis plateau that shows the largest concentration of hot spot volcanism.

Since there are hot spots on Venus and new lava is added to the crust, it follows that there should also be zones of downwelling on the planet, where older crust is destroyed to balance out the flux of matter at the surface. The downwelling areas of Venus could be the crumpled plains (tessera terrain) where fissures are sealed shut and volcanic centers are rare.

In summary, Venus today is a planet dominated by "spotty" hot spot volcanism, without the intricate network of rifts and trenches that on Earth cut up the lithosphere into well-defined plates. But was this always the case? Is this a permanent and stable configuration, or is it only the most recent phase of a much more complex geological history?

THE LIFE AND TIMES OF VENUS

We have so far skirted the issue of time in our survey of Venus volcanoes. Is volcanism ongoing today? What are the age estimates of the major edifices, the rate of lava emplacement at the surface? Because its volume is 85% that of the Earth, Venus possessed an equivalent proportion of radiogenic elements and should have been able to sustain volcanism over a long period and arguably up to this day.

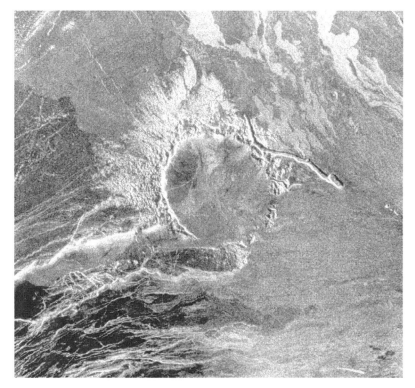

FIGURE 7.17 The chronology of geological events on Venus is recorded
in overlapping stratigraphy. Here, a 63-km impact crater is partially
overrun by lava that moved in from the east. The crater, named Alcott,
is located in Lavinia Regio (59.5° S, 354.5° E) Credit: NASA/JPL-Caltech,
courtesy of Brown University.

The youthful appearance of many volcanoes on Venus, albeit
erosion is much weaker on our sister planet, speaks in favor of recent,
if not contemporary volcanism. Short of Venusian lava samples that
we can date in our labs, age estimates are limited to the tally of impact
craters on the surface. The task is complicated by the fact that the
thick atmosphere destroys the smallest, most numerous projectiles.
There are no impact craters on Venus smaller than 20 km in size and
their total number barely reaches one thousand. Hence, their average
density is about two impact scars per million square kilometers,
which yields an age estimate of roughly 800 million years for the

planet's surface as a whole (the estimate ranges from 300 million years to one billion years depending on the model). Relative to this planet-wide average, some geological units such as the crumpled tessera terrain appear slightly older, while most large volcanoes and lava fields display only half the crater density and should be half as old. There are also very young volcanoes on Venus, like Maat Mons, that cover even the freshest impact craters with lava. These volcanoes might be younger than 100 million years, leaving open the possibility that some might even be active today.

Although it is logical to assume that contemporary eruptions do take place, we have still to catch a volcano in the act, a difficult proposition if we consider the eruption rates postulated for Venus. Dividing the volume of lava provinces on Venus by their estimated age bracket yields a variety of output rates. The voluminous lava plains of about 500 million to one billion years ago were emplaced at an average rate of 1 to 3 cubic kilometers per year, which is comparable to present-day volcanism on Earth at the mid-ocean ridges. However, the younger volcanic edifices of Venus – those lavas emplaced in the last 500 million years or so, represent much less volume. Calculated rates are correspondingly low: probably little more than $0.1 \, km^3$ of lava per year, which is 20 times less than the present volcanic output of the Earth. In fact, the average figure for Venus is comparable to that of a single volcano on Earth, such as Kilauea or Mount Etna. Moreover, this estimated figure is an average. At such a rate, Venus could be quiet for a century, erupt $50 \, km^3$ in less than a year and go quiet again for a century. Or the planet could stay dormant for millions of years at a time, then emplace one great lava field in a matter of weeks or months, as it apparently did at Mylitta Fluctus (see our field trip in the next chapter). The fact that no eruption was observed during *Magellan*'s five-year mapping mission should come as no surprise.

One way to explain the average surface age of 800 million years on Venus is that a major event happened at that time, renewing the entire crust – resetting the clock, as it were. In contrast to the Earth, where plate tectonics has been churning out the heat of the interior at a steady pace, Venus might undergo an oscillatory regime. The interior heats up until a catastrophic "overturn" takes place, the dense crust foundering into the hot mantle to be replaced by a new one. The new crust cools and thickens in turn over hundreds of millions of years, storing up the mantle's heat until the next overturn is triggered.

In this model, the last overturn took place around 800 million years ago, with a flurry of volcanic activity that is still visible today. Since this global resurfacing event, volcanism switched to a trickle, building up the individual volcanoes at a snail's pace as we await the next crisis.

Other models suggest that this switch of regime took place only once, 800 million years ago. Before that time, the mantle of Venus was hot and fluid. It was supplied with energy from the molten core and abundant volcanism took place at the surface, perhaps in the form of plate tectonics. The core then froze solid. The mantle soon expended its excess heat and slowed down its convection to a crawl. Since this turning point, 800 million years ago, volcanism remains at the low level that we witness today.

MELTING THE SURFACE

If it occurred, the global volcanic event that resurfaced Venus certainly had a major impact on the environment. The eruption of over a billion cubic kilometers of lava – one thousand times the size of an entire volcanic province on Earth – certainly released huge amounts of volcanic gases in the atmosphere. These were principally carbon dioxide, water vapor and sulfur dioxide. Because the concentration of carbon dioxide in the atmosphere was already enormous, the addition brought little change to the greenhouse effect. On the other hand, the background amount of water vapor and sulfur dioxide was probably low when the volcanic event took place, so that their respective figures might have jumped ten-fold and one hundred-fold in the process.

One consequence of such a jump in water and sulfur content

FIGURE 7.18 Sif Mons is a 300-km-wide volcano, with a 50-km-wide, radar-dark summit caldera. Bright lava flows streak the lower flanks. Credit: NASA/JPL-Caltech, courtesy of Brown University.

might have been the thickening of the cloud cover, reflecting more sunlight into space and lowering the atmospheric and surface temperatures by up to 100 K (100 °C). Over time, the extra sulfur would have reacted with minerals on the ground, so that the sulfuric clouds were eliminated on a time scale of one hundred million years, but the extra water vapor lingered on, boosting the greenhouse effect. Hence the initial cooling was followed by a swing in the opposite direction, with temperatures peaking 100 K above today's figures. Finally, the progressive bleeding of water vapor off into space brought the temperature back down to the current figure.

This is only a model, but this temperature swing might explain some of the geological features that we observe on the surface of Venus. The crumpled tessera plateaus could be explained if the crust and upper mantle of Venus had once been 100 K hotter than today, causing it to expand and then contract and crumple as the temperature returned to normal. Likewise, the putative carbonate sediments at the surface, dangerously close to their melting temperature, might have "gone over the edge" during the heat pulse and liquefied catastrophically, gushing like motor oil and creating the long channels that so mystify volcanologists today.

Venus might then be a showcase example of how volcanic gases not only affect the atmosphere of a planet but also shape its surface through a greenhouse effect that reaches deep down into the crust. The understanding of Venus' volcanism is a fundamental challenge that beckons us to send more probes to explore the surface. Given the hellish temperatures and pressures, the landing of crews might forever be off limits, but we can at least imagine what it would be like to drift over the surface in futuristic aircraft to visit our favorite sites. It is this guided tour of Venusian volcanoes that we propose in the next chapter.

8   A tour of Venusian volcanoes

Theia Mons (Beta Regio)

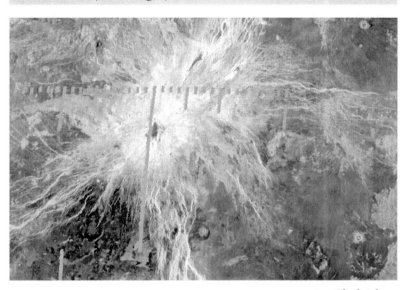

Latitude:	20° N
Longitude:	276° E
Altitude:	4000 m
Width (lava field):	800 km
Caldera:	75 km × 50 km
Age:	<100 million years?

FIGURE 8.1 The bright star-like pattern of Theia Mons straddles several rift zones in Beta Regio. A dark elongate caldera tops the volcano, alongside a gray strip where data were lost. *Magellan* radar image. Credit: NASA/JPL-Caltech.

For our first landfall on Venus, it is fitting that we explore Beta Regio – the first volcanic region identified from Earth by radar, which proved to be one of the most spectacular rift zones on our sister planet.

Beta Regio is a volcanic rise located in the northern hemisphere, between 15° N and 35° N, and 275° E and 290° E. It comprises two large edifices: Rhea Mons to the north and Theia Mons to the south.

Theia Mons is the taller of the pair (4000 m) and lies at the intersection of three major rift zones. The northern rift, Devana Chasma,

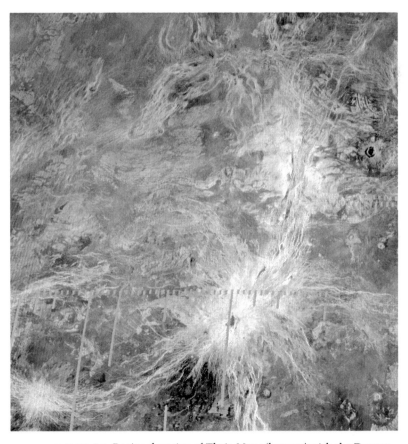

FIGURE 8.2 Regional setting of Theia Mons (bottom) with the Devana Chasma rift zone extending north up through the top bright massif of Rhea Mons. Credit: NASA/JPL-Caltech.

is up to 240 km wide and extends 800 km to Rhea Mons, splitting the latter in two. Beta Regio has been compared to the East African Rift, on account of its topography and branching fault zones. Like the East African Rift, it probably owes its origin to a hot spot, whereby a plume of buoyant, partially molten material uplifts and stretches the crust, sending offshoots of magma that build up volcanoes at the surface.

Theia Mons is one of the largest shield volcanoes on Venus, with lava flows radiating hundreds of kilometers in all directions. The flows partially fill Devana Chasma and are also sliced up by its faults – proof that rifting and volcanism are happening at the same time. An oval-shaped caldera, over 50 km wide, crowns the volcano. It was probably formed by the collapse of the summit as magma traveled underground to the flanking rift zones.

Because of the sharpness of its features, Theia Mons is considered to be very young (less than 100 million years) and might still be active today.

Itinerary

Landing on Venus is a dangerous proposition, but we can imagine that manned spacecraft will, one day, rise to the challenge. Pressures at the surface of Venus are comparable to those 1000 m under the sea here on Earth, and the temperature is a whopping 735 K (460 °C)! The Soviet probes of the 1970s survived less than one hour on the ground, but we can hope that our manned spacecraft will withstand several days of such harsh treatment. The trick will be to frequently retreat to the lower temperatures and pressures of the upper atmosphere, between short expeditions on the ground, perhaps using a balloon system.

To begin our journey, we would want to land on the eastern slope of the Beta rise, as did the first spacecraft to take a picture of the surface of Venus (*Venera 9* accomplished this feat on October 16, 1975). We attempt to land alongside the Russian probe, in the same way that the crew of *Apollo 12* joined the unmanned *Surveyor 3* craft on the Moon (see Fig. 3.3).

BEHEPA-9 22.10.1975   ОБРАБОТКА ИППИ АН СССР 28.2.1976

FIGURE 8.3 *Venera 9*, the first probe to take an image of the ground of Venus landed east of Beta Regio. Credit: Russian Academy of Science, courtesy of NSSDC, Goddard Space Flight Center.

In our spacecraft, we plummet through the cloud deck of Venus and emerge at an altitude of 30 km to discover the landscape below, through the crystal-clear atmosphere of carbon dioxide. On the western horizon, we contemplate the large shield volcano Rhea Mons, sliced open by the Devana Chasma rift valley. We descend towards the lava plains at the foot of the volcano, gouged by faults that resemble giant runways. We spot *Venera 9* at last: a ball of shimmering metal, precariously perched on a steep slope. Around the spacecraft, flat-topped "tiles" of lava, up to a meter across, rest on a talus of finer soil.

Once safely on the ground, we take in the scenery through the porthole. The ground appears orange, with no shadows. Indeed, one Russian scientist has compared the experience on the surface to "being inside a giant fluorescent light bulb." Some of the rocks even appear to glow like embers. No wonder, given the oven-like temperature outdoors!

We want to extend a grappling arm to remove part of the *Venera* spacecraft and return it to Earth. Its analysis back at the lab would reveal how much corrosion the robot suffered, during so many decades spent in the planet's acid oven.

We also want to sample the lava rock. According to the initial analysis transmitted by *Venera 9*, back in 1975, we know that the lava is relatively rich in potassium (0.9%). This would place it in the family of alkaline basalts, which are common in rift valleys on Earth. The moderate potassium content, in particular, is reminiscent of lava in the East African Rift, to which Beta Regio has aptly been compared.

After this first sampling stop, we rise from the fiery surface of Venus and retreat to the relative coolness of the upper atmosphere, below the cloud deck, from where we keep a clear view of the ground. Drifting west, up the Beta rise, we soon reach the edge of Devana Chasma, stretching north to south across the volcano. We descend into the broad rift valley and follow it to the south, scrutinizing fractures and fissures below our flight path. The western scarp of Devana Chasma, in particular, is especially steep and its golden flanks shimmer in the diffuse daylight. It is difficult to make out if the massive swell of Rhea Mons, in the background, is truly a volcano. In any event, it looks ancient, with a crumpled surface and the overprint of Devana's recent rifting. We would want to take a sample of its surface, to determine if it is indeed an old volcanic edifice that dates back to the early history of Beta Regio.

Next, as we head south towards Theia Mons, comes the spectacular Somerville impact crater, 37 km in diameter, lying in the

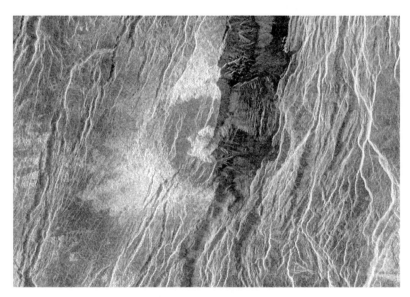

FIGURE 8.4 The Balch (also known as Somerville) impact crater, 37 km diameter, is split in two by the Devana rift zone, which is slowly stretching the crust in an east–west direction. Credit: NASA/JPL-Caltech.

middle of the rift zone. The circular depression is cut straight through by fresh fractures that are clearly splitting the crater in two halves. Peering through our portholes, we estimate the east–west extension across the crater floor to reach 10 km. After Somerville crater, our attention turns to the wide shield of Theia Mons, now looming ahead. Devana Chasma splits the northern flank of the volcano wide open, in the same way that the wedge-shaped Valle del Bove digs into the flank of Mount Etna.

On Theia Mons, the rift is 50 km wide and 3 km deep. Its bounding scarps slice through the lava pile, revealing millions of years of edifice construction. Sending a robot down the scarp to analyze the layers would be a fine objective for a future mission.

The rift narrows towards the summit, until it is completely covered up by lava flows. At the peak altitude of 4000 m, the shield is crowned by an oval caldera, 75 km long and 50 km wide, that compares in size with the giant features atop Martian volcanoes. Theia's caldera is covered by a radar-dark formation. As we lower our balloon towards the crater floor, we crowd around the narrow portholes to get our first glimpse of the surface. Is it an ancient lava lake? Or else dunes of pulverized ash?

The future will tell.

Sapas Mons

FIGURE 8.5 Perspective view of Sapas Mons, with the Aphrodite rift zone on the horizon. Credit: NASA/JPL-Caltech, courtesy of Brown University.

Latitude:	8° N
Longitude:	188° E
Diameter:	600 km
Altitude:	3600 m
Relief:	2400 m
Volume:	30 000 km³
Slope:	0.3° to 11°
Caldera:	100 km
Depth:	<500 m
Age:	50 million years?

Sapas Mons is typical of large volcanoes on Venus. It is a broad shield, 600 km in diameter, that reaches a mere 2400 m above the surrounding plains. This gentle relief limits the volume of the volcano to 30 000 km³, comparable to that of a large Hawaiian volcano. It features a complex summit with two peaks, which makes it one of the most photogenic landscapes on Venus.

Sapas Mons is located in Atla Regio, a volcanic rise at the eastern end of Aphrodite Terra that features branching rift zones. Two other large volcanoes rise to the southeast: Ozza Mons and Maat Mons, which straddle the rift zone.

Sapas Mons sits atop its own topographic swell and is coated with abundant, radar-dark lava flows. The swell is typical of hot spot volcanism, in which a plume of magma lifts and stretches the crust, and erupts voluminous amounts of lava at the surface (the dark plains). The tail end of the plume builds a more restricted pile of lava: the edifice proper. The swell is slightly asymmetrical and the lava spreads mostly to the northwest and southeast, resulting in a butterfly pattern.

The radar brightness of the lava fields change from base to summit, dividing the edifice into six major units. These probably

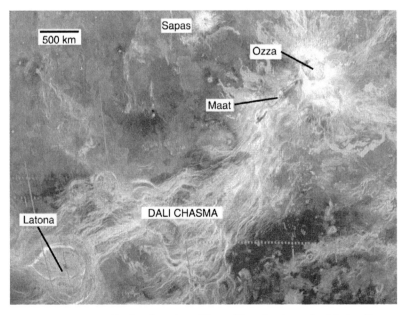

FIGURE 8.6 Regional setting of Sapas Mons (top), north of Aphrodite Terra and its volcanoes Ozza and Maat Montes. Credit: NASA/JPL-Caltech, courtesy of LPI (slide set: *It's a Dry Heat: the Geology of Venus from Magellan*, compiled by Robert R. Herrick and Maribeth H. Price).

reflect different batches of magma flowing down the edifice, with the longest and oldest ones cropping out on the periphery of the volcano, and the younger eruptions covering the substratum with progressively smaller flow fields. The radar properties of the different units suggest different textures. For example, unit 2 flows, on the lower flanks, are made up of narrow, sinuous flows, with finger-like appendices that suggest ropy pahoehoe lava. On the other hand, units 5 and 6 near the summit may be composed of rougher, blocky aa flows.

Lineaments streak the lower flank in several places (especially to the north and south) and are associated with small cones and domes. Such lineaments are probably dikes of magma that propagate away from the central magma chamber and vent small amounts of ash or lava at the surface.

Towards the summit, ring faults define a subdued caldera, 100 kilometers in diameter, which betrays the size and location of the underlying magma chamber. Twin domes, each 25 km in diameter and a few hundred meters in height, decorate the summit. Their steep flanks are fluted by mass wasting into a stellar array of ridges and valleys. The valley fill is radar-dark and probably composed of fine ash. Short stubby flows surround the domes and might be made of siliceous or volatile-rich lava that was last to ooze out from the cooling magma chamber.

The age of Sapas Mons, like the age of all volcanoes on Venus, is poorly constrained. Sapas is younger than the impact crater to the east (embayed by its lava flows), but older than neighboring Maat Mons, which is probably one of the youngest volcanoes on Venus. We guess Sapas Mons to be around 50 million years old.

Itinerary

As for all our Venus field trips, it is wise for us to stay inside a pressurized "diving bell" similar to deep-ocean submarines, stare out at the landscape through thick portholes and collect samples with a robotic arm. We use a balloon to lift and transport our capsule to various points of interest up the volcano.

We approach Sapas Mons from the east, flying over the Atla rift zone. The volcano looms large on the horizon, bathed in a diffuse golden light. The base of Sapas Mons is defined by a jagged contact of sinuous lava flows, spread out like giant fingers atop the darker plains. One flow abuts a large impact crater, 25 km in diameter, and splits into two branches that bypass the obstacle north and south. The fact that the flow did not overrun the rim of the crater suggests that it is relatively thin.

We follow the margin of the edifice counter-clockwise to the north, heading for an array of bright fractures that streak across the lower flanks. As our balloon lowers us to the ground, we notice a cluster of small "parasitic" shields that average 3 to 6 km in diameter and a few hundred meters in height, straddling the fracture zone. We land atop the gentle slope of the easternmost shield that displays a small summit pit. The close relationship between these parasitic shields and the fractures hint that the latter might be dikes of magma that congealed underground. Offshoots of magma that reached the surface built up the shields. If this interpretation is correct, the samples grabbed by the robotic arm could be small flows or welded scoria that spattered and fused together as they fell around the vent.

Our examination complete, we lift off from this first site and head up the fracture zone towards the summit. We proceed across the upper flows, known as "unit 5" in the original description by Susan Keddie and Jim Head. This is a field of radar-bright flows that are broader and less sinuous than those lower down. From the radar signal, they appear to be rough on a small scale – perhaps rubble-like lava of the aa type. Through the portholes, we strain to make out the texture of the land below. There are no channels, no give-away details of any kind. Perhaps the lava fields were emplaced as great crusted-over sheets, with the magma flowing and bulging underneath the surface (inflated flows).

A mere 50 km from the summit, arcuate fractures slice through the lava flows. This set of ring faults mark the boundary of a wide subdued caldera. The first fault we cross is enlarged by a series of pits, a couple of kilometers wide. Other rings crack the surface and accommodate the collapse of the volcano summit, in

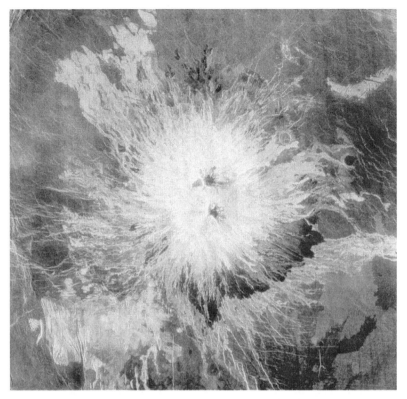

FIGURE 8.7 Plan view of Sapas Mons with its double summit and contrasting lava flows. See map for geological interpretation. Credit: NASA/JPL-Caltech.

response to the underground shifting of the magma. Since the diameter of a caldera mirrors that of the underground reservoir, the latter appears to be about 100 km in diameter.

Within the central area, two domes shimmer in the golden light. They constitute Sapas Mons' trademark double summit. Each dome is roughly 25 km wide, with a flat-looking top and fluted slopes – alternating ridges and valleys that carve up the flanks. Since there is no water erosion on Venus, this scalloped appearance is probably caused by landslides, triggered by gas-rich eruptions. Radar-dark streaks that stream down the valleys might be avalanches of smooth ash. The disruption of magma into pulverized ash is indeed more

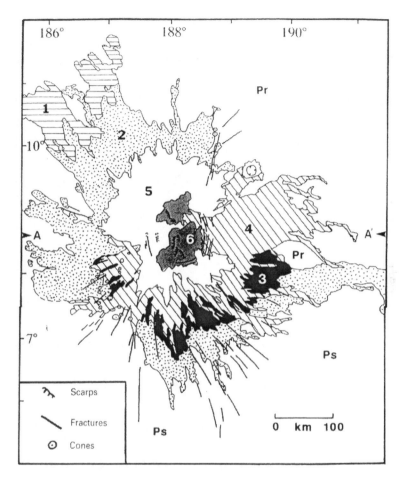

FIGURE 8.8 Map of Sapas Mons. The lava cover is subdivided into six units from oldest in the northwest (1) to youngest near the top (6). Parasitic cones are concentrated on the southwestern flank. Credit: from S. T. Keddie and J. W. Head, Sapas Mons, Venus: evolution of a large shield volcano, *Earth Moon and Planets*, **65**, 129–90.

probable at the summit of a tall volcano like Sapas Mons (3600 m above the reference level), since the pressure at the summit falls below 8 MPa (80 bars) – which is 20% less than at the foot of the volcano. Bubbles are less constrained by the pressure, expand to larger sizes and have a better chance of disrupting the magma. Explosive behavior at the summit of Sapas Mons is all the more probable

FIGURE 8.9 The double summit of Sapas Mons, surrounded by bright lava flows. Credit: NASA/JPL-Caltech.

because magma in the upper part of the volcano might be differentiated, siliceous and volatile-rich. This would explain the viscous-looking flows that surround the summit domes (unit 6).

In order to test this hypothesis, we aim for a landing spot at the foot of the northern dome. A polygon-shaped mesa, surrounded by spiny ridges, it has the appearance of a giant insect (actually, the modified domes of Venus were nicknamed "ticks" when they were first discovered).

A steep scarp cuts off the western side of the mesa and we steer our vessel to land at its base, where a sharp contact separates radar-dark deposits to the south from brighter hummocky lava flows to the north. From the safety of our diving bell, we send a robot into the blazing environment to collect samples of both these units, and to take photographs of the scarps to document any layering. As soon as this brief reconnaissance is complete, we take off and pull out of the inferno, quickly ascending into the cooler reaches of the upper atmosphere. As we reach the cloud deck, we pause a minute to catch one last bird's eye view of Sapas Mons.

Gula Mons and Idem-Kuva Corona

FIGURE 8.10 Gula Mons (left) with lava flows spilling down slope over Idem-Kuva Corona (center). Another volcano, Sif Mons, rises in the background (right). Credit: NASA/JPL-Caltech, courtesy of Brown University.

Gula Mons

Latitude:	22°N
Longitude:	359°E
Dimensions:	400 km × 250 km
Altitude:	3200 m
Relief:	1700 m
Caldera:	double (35 km and 25 km)
Age:	>400 million years

Idem-Kuva

Latitude:	25°N
Longitude:	358°E
Diameter:	225 km
Relief:	600 m
Age:	>400 million years

Gula Mons is a relatively steep volcano that is superposed on the broad topographic swell of Eistla Regio, in the northern mid-latitudes of Venus. Gula Mons is connected to the north, by way of a rift zone, with an exotic volcanic landform: Idem-Kuva Corona. We will examine both features in this field trip.

Gula Mons is a typical shield volcano that grew at the intersection of two major faults, on the "hot spot" swell of Eistla Regio. The basal apron of lava reaches an overall diameter of 1200 km (it would bury France or Texas) and its average slope is less than one degree. The superposed central part of the construct is steeper (1° to 5° slopes), with a diameter of roughly 300 km. This central section is made up of narrow, finger-like flows, as opposed to the more voluminous lava fields at the base.

Gula Mons has an elongate summit, torn by a NE–SW trending rift, 150 km long and 30 km wide. At each end, there is a filled-in caldera. The southwestern one reaches 35 km in diameter, and the northeastern one 25 km. Extending from the latter, a rift rips down the northern flank of the volcano and onto the surrounding plateau, where it merges with the volcanic complex of Idem-Kuva.

FIGURE 8.11 Plan view of Gula Mons (dumbbell-shaped bright summit, center) and Idem-Kuva Corona (right) with its two bright lava flows extending towards edge of frame. North is to the right. Credit: NASA/JPL-Caltech.

Named after the Finnish spirit of harvest, Idem-Kuva is a corona, a particular class of Venusian volcanoes. It is characterized by a circular dome or plateau, surrounded by a moat, which gives it the shape of a cowboy hat. The central plateau is 100 km wide, but rises only 600 m. The peripheral moat is bounded by ring faults that are prominent in the south and buried by younger lava in the north. Two major flows, which appear bright in radar imagery, emerge from the ring faults on either side of the corona, like ribbons from a hat.

Apparently, Gula volcano and Idem-Kuva Corona are two related forms of magmatic activity on Venus, the former being the extrusive manifestation of a hot spot and the latter being its intrusive, underground version. Both Gula and Idem-Kuva are relatively ancient landforms, since they are blanketed by the ejecta of several large impact craters. Their age probably exceeds 400 million years. Gula Mons appears to post-date Idem-Kuva since its lava flows are deflected around the corona.

Itinerary

We approach the broad swell of Eistla Regio from the west, in order to fly over an additional volcano – Sif Mons – on our way to our target, Gula Mons. Sif Mons is 700 km west of Gula, but their extensive lava shields nearly overlap. Seen from our balloon, Sif Mons shows a spectacular rifted summit and a caldera 50 km wide, filled with smooth lava flows that breach the eastern rim and flow down the flank of the volcano. We take in the view before setting our sights on a second shield in the distance: Gula Mons.

As we lose altitude, the tall volcano begins to show its bumpy profile. There is a noticeable change of slope halfway up the mountain. The steeper upper half ends in an elongated summit, torn lengthwise by a rift zone. Lava flows coat the summit area and cascade down the slopes.

We aim for a landing spot at the northern extremity of the rift, where it nests a caldera 25 km in diameter. The crew on board our

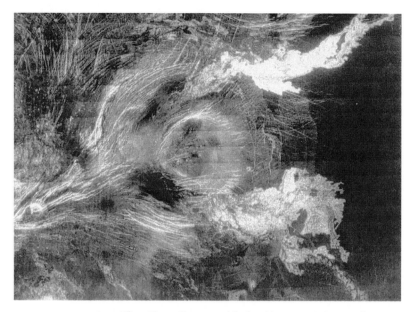

FIGURE 8.12 Idem-Kuva Corona, with the rift zone originating from Gula Mons to the left, and two bright lava fields flowing out of the corona to the right. North is to the right. For perspective view, see Fig. 8.13. Credit: NASA/JPL-Caltech.

diving vessel makes final preparations for the short touchdown in the searing heat in order to quickly grab rock, soil and gas samples.

Radar surveys show the summit of Gula Mons to be exceptionally bright, like most elevated regions on Venus. This implies unusual electrical properties of minerals at the surface. One hypothesis is that the slightly lower temperature and pressure at these elevations favor the reaction of sulfurous gases with the lava. This chemical weathering could yield sulfide minerals such as *pyrrhotite* and *pyrite* ("fool's gold"). Our sampling stop on Gula Mons will hopefully solve this mystery.

With our samples stowed away, we lift off the summit and glide down the northern slope of Gula Mons over the troughs and faults that betray the continuation of the rift zone across the plateau. The lineaments lead us directly into Idem-Kuva, a prime example of the unique corona family. Idem-Kuva is a circular structure consisting of

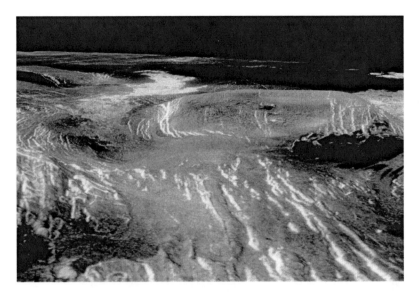

FIGURE 8.13 Perspective view of Idem-Kuva Corona, looking north, with the rift zone in the foreground and one of the bright lava flows in the distance. Credit: NASA/JPL-Caltech.

a central lens of uplifted terrain – 100 km wide and a mere 600 m in relief – surrounded by a moat of lower elevation. Fractures lead into the complex and slice radially through the center. Two other sets of lineaments deflect around the margins of the corona, on either side. These western and eastern fracture belts switch into fans of bright lava flows that extend northward for another 200 to 300 km.

Coronae are believed to be the surface manifestation of magma plumes that stall under the crust and spread laterally without breaking to the surface. Are the bright lava fields at Idem-Kuva an exceptional break-through out of the corona? Or do they come instead from nearby Gula Mons, through underground pathways along its northern rift zone?

We land our vessel at the source of one of the bright lava fields, in order to drive to the central uplift with our pressurized rover. A close-up look at the uplift would tell if it were older crust, inflated and uplifted by an underground plume. As for the bright lava flow, collecting and analyzing a sample will tell us if the lava comes directly from the mantle plume under the corona, or if it decanted in a magma chamber under Gula Mons.

Pancake domes (Alpha Regio)

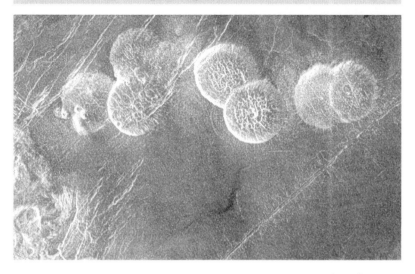

Latitude:	30° S
Longitude:	11.8° E
Number of domes:	7
Average diameter:	25 km
Average elevation:	750 m
Average slope:	up to 30°
Age:	undetermined

FIGURE 8.14 Seven pancake domes, east of the Alpha Regio plateau. Each is roughly 25 km across. Credit: NASA/JPL-Caltech, courtesy of Brown University.

Steep-sided domes constitute one of the oddest categories of volcanoes on Venus. They are nicknamed "pancake domes" in view of their circular plan form and flat summit, and their origin is much debated. There are approximately 150 such domes on Venus, many of which occur in clusters. Some are associated with coronae, while others lie along the margins of ancient plateaus of crumpled crust

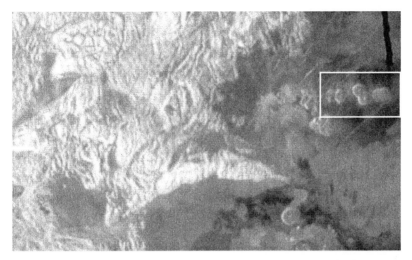

FIGURE 8.15 Regional view of eastern Alpha Regio. Pancake domes are on right edge of frame. Credit: NASA/JPL-Caltech.

(*tesserae*). They average 20 to 25 km in diameter (the largest ones reach 100 km) and their relief is on the order of a few hundred meters.

One of the most spectacular examples of pancake domes is a chain of seven features on the eastern edge of Alpha Regio. All seven are equal in size (25 km in diameter, 700 m in height) and most are grouped in overlapping pairs. Their broad summits are cracked by fractures – some radial, some concentric – and display small pit craters. The westernmost dome also features a cluster of cones on its flank.

The symmetry and cracked surface of pancake domes led geologists to propose an original mode of formation, which has them grow as single "blobs" of viscous lava that spread out radially in all directions. On Earth, such *endogenic domes* are associated with siliceous magma (usually dacite or rhyolite). Their solid surface is stretched and cracked as the magma intrudes and "inflates" them. Avalanches of hot debris (pyroclastic flows) often occur along their slopes.

On Venus, the association of pancake domes with ancient crust supports the notion that siliceous magma is involved. Heating of the thick crust by a mantle plume might lead to the melting of its most fusible, silica-rich fraction. On the other hand, the high viscosity of pancake domes might be caused by abundant gas dissolved in the magma. The high pressure at the surface of Venus would prevent the emulsion from blowing up. It would spread instead in all directions, as one very thick flow. Radial cracks record the expansion of the lava pancake, while concentric cracks testify to deflation and collapse as it later cools and subsides.

Regional fractures strike northwest across the plains – some buried by the domes, while others cut straight through them. This dual relationship indicates that the build-up of the domes was synchronous with the fracturing of the plains, which opened up pathways for the magma to reach the surface. Although the pancake domes are certainly younger than the ancient plateau of Alpha Regio, their exact age is difficult to determine. A sampling expedition might be necessary.

Itinerary
Our airship flies eastward across the vast plateau of Alpha Regio on its final approach for landing, at the foot of the seven domes. The crumpled terrain resembles an old mountain range – an endless succession of ridges and troughs – and must likewise represent on Venus a thicker crust than average. Mantle plumes are unable to break through such a thick lid. There is no major sign of volcanism at the surface, except for a few breached cones and lava flows on the plateau's margin.

We cross the eastern edge of Alpha Regio, its cliff line embayed by smooth lava plains that stretch out to the horizon. Our target is coming up 200 km ahead, and we start lowering our altitude. A breached cone stands out above the plains, then a cluster of small domes and vents, signaling the beginning of the volcanic province.

The first pancake dome looms ahead – the least regular of the bunch with three cones growing on its southwestern flank. We fly

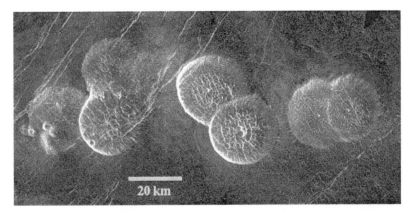

FIGURE 8.16 The seven pancake domes: our landing site is in the
narrow pass between the two pairs at right. Credit: NASA/JPL-Caltech.

directly over the "parasitic" cones that look surprisingly like terrestrial volcanoes. They average three kilometers across with steep
flanks and a wide central crater, like Vulcano in Sicily and craters in
the Galapagos Islands.

The second pancake dome overlaps the first one along a perfectly circular – and steep – flow front. The symmetry of the edifice
looks impressive as we fly across the top. The 25-km-wide lens of lava
is cracked by radial fractures, running from center to edge. The radial
deformation pattern hints that the volcano puffed up as a large mass.
There is no pit crater at the center, but we notice a small crater on the
southwestern flank. A couple of pits are also visible to the north, at
the junction with a third dome. As for the eastern flank under our
flight path, it is torn by a regional fault that continues down into the
plains.

We cross a short stretch of lava plains before reaching the next
pair of overlapping pancake domes and fly over the southern one. This
time, the fractures at the summit are both radial and concentric, in a
spider web pattern, attesting to both inflation and deflation in the life
of the volcano. At the center of the pancake lie two coalesced pits –
the size of Kilauea caldera in Hawaii. We fly over the shallow collapse
bowl and down the eastern flank of the volcano.

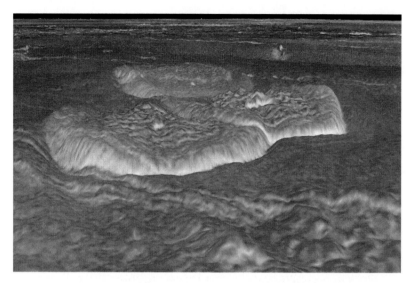

FIGURE 8.17 Perspective view of pancake domes, looking to the northeast. The vertical exaggeration is 23×. Credit: NASA/JPL-Caltech, JPL Multimission Image Processing Laboratory.

Our airship manages a smooth landing at the foot of the dome, in the narrow pass that separates it from the last pair of pancake domes. From this spot we can reach the volcanoes on both sides – less than five kilometers apart – and study the debris flows that are shed off the slopes.

We will dispatch robots that will drive up to the closest dome, in the oppressive heat, and explore its slope with tools and a video camera. We will be able to pick out a freshly broken piece of lava from the steep flow front. Will it be silica-rich lava, such as andesite or trachyte, seldom found on planets outside Earth? Or will it be ordinary basalt, puffed up into a glassy froth by an overabundance of gas?

As for the study of ash falls and pyroclastic flows on the site, we can directly drill the ground under our airship to collect a deep core. Once the heist is complete, we must waste no time, lift off the red-hot surface of Venus, and head up to the clouds.

Ammavaru and Mylitta Fluctus

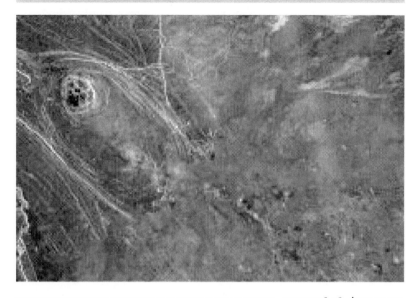

Ammavaru

Latitude:	47° S
Longitude:	20° E
Dimensions:	400 km
Caldera:	40 km × 20 km
Age:	<50 million years

Mylitta Fluctus

Channel length:	1200 km
Channel width:	5 km
Field area:	>100000 km^2

FIGURE 8.18 Ammavaru volcano, surrounded by an oval set of faults, is the source of an extensive lava field. Credit: NASA/JPL-Caltech.

This field trip is a cruise down a great "river" of Venus: a lava channel that winds 1200 km from its source volcano (Ammavaru) to a giant lava field (Mylitta Fluctus). The volcano is modest in size but its lava field covers 100000 km^2 – the area of a flood basalt province on Earth, like the Columbia River plateau.

The tectonic setting here is different than others we have so far encountered on Venus. Rather than a large domical uplift, this volcanic province in the southern hemisphere is a linear chain of coronae, stellate centers and volcanoes, hugging the boundary between the highlands (Lada Terra) and the plains (Lavinia Planitia). The linear rift first strikes eastward (Ammavaru is located on this segment), before heading north towards Alpha Regio.

It is unclear whether deep mantle plumes initiated the rifting, or if the rift came first and was then infiltrated by hot spots. Ammavaru volcano and its lava field are not deformed by the rift and are therefore younger. They are late events in the history of the region, and we guess them to be less than 50 million years old.

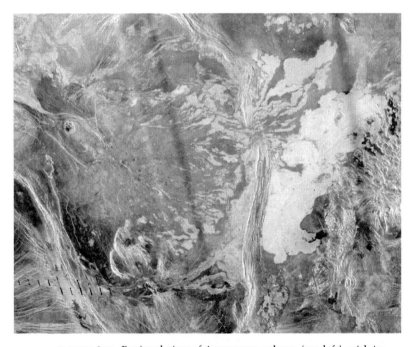

FIGURE 8.19 Regional view of Ammavaru volcano (top left), with its main lava channel emerging from a cleft on the left margin of the image and heading south. The channel then veers east (bottom of image), fans out into three branches (bottom center) and creeps up along a ridge belt, before breaking through a pass (top right) and spreading out as a wide bright flow field (right side of image). Credit: NASA/JPL-Caltech.

Ammavaru volcano is an oval-shaped collapse caldera, girdled by ring faults. The central depression is 200 km long (in a NW–SE direction) and 100 km wide. In the northwestern corner of the depression sits a nested caldera, roughly 40 km in diameter. Outside the depression, the flanks of the volcano rise to an elevation of 500 m and define a broad shield, 400 km across, with lava flows fanning out in many directions.

The channel that leads out to the grand lava field of Mylitta Fluctus originates on the southwestern flank of the volcano. It emerges from a collapse feature and follows a broad U-shaped course, first flowing south, then east, before returning north along a ridge belt. The channel finally breaches the belt and dumps the lava along its eastern front as a swirling pool of radar-bright flows, more than 100 000 km^2 in area.

The channel, 5 km wide at its source, runs a total length of 1200 kilometers, comparable to that of the Snake River in the United States or the Rhine River in Europe. It takes on a variety of shapes, beginning as a straight rille, spreading into a braided network, and finally splitting into distributory branches.

Geologists recognize six episodes of lava flows around Ammavaru volcano, beginning with the construction of the shield proper. The last episode is the outpouring of lava on both sides of the ridge belt (Mylitta Fluctus). According to experts, these voluminous flows were emplaced in a few months at most, as hot, fluid and turbulent magma of basalt or komatiite composition.

Itinerary

To begin our "cruise," we drop out of the clouds straight over the Ammavaru lava shield and its central fault-rimmed depression. Our bird's eye view reveals a myriad of fractures at all scales. On the southern flank of the edifice, parallel to a fracture swarm, a collapse feature marks the source of the lava channel. Its tadpole motif, complete with dendritic "legs," extends over a length of 50 km before pinching out into a narrow outlet that follows a southeast-trending

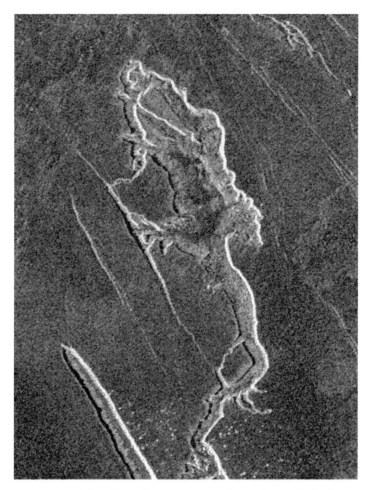

FIGURE 8.20 A box-like collapse source, on the southwest flank of
Ammavaru volcano, merges into a 5-km-wide lava channel that heads
south, then east to feed the Mylitta Fluctus lava field. Credit:
NASA/JPL-Caltech, NSSDC/GSFC.

trough. Along this segment, the main channel is 5 km wide and displays a central lava-filled valley, flanked by radar-bright levees.

We drift southward down the gentle slope of the channel, which
follows a ridge belt on its right side and lava plains on its left. About
300 km downstream from its source, the channel is flush with the
lava plains and opens up into a broad network of braided channels as

it picks its way eastward. The channels divide up around streamlined islands – last knobs of terrain left standing after the carving and plucking of the turbulent lava flow. The channel branches snake across a wide flood plain of lava, before reuniting to cross a narrow pass between spurs of ridged terrain.

Out of the pass, the lava river spreads out into a distributory network – a bird-foot delta. Three channels diverge, two striking east and one cutting to the northwest, each surrounded by a field of radar-bright lava.

We continue our flight along the northern branch and ponder over the conditions that allowed lava to flow over such a great distance. High temperature? Low viscosity? Extraordinary discharge rate?

According to our flight instruments we are 600 km downstream from the source and the channel shows no sign of ending as branches unite and strike north, hugging the front of a major ridge belt. Here we begin to embrace the extent of the lava flows that spread along the ridge, as we proceed northward for another 300 km. We are reaching the 1000-km mark on our flight path, and at our modest speed of 50 km/hour in the sluggish Venus winds, the clock has been ticking for nearly 24 hours. No one even thinks of sleeping as we approach the spectacular breach in the ridge belt, through which the lava channel threads its way eastward.

Emerging from the pass in the ridge belt, the channel opens up on a huge expanse of lava: Mylitta Fluctus. The radar-bright flows fan out in all directions, as far as the horizon. They extend 100 km eastward and more than 300 km north and south, covering an area larger than 100000 km².

We drift upward in our airship to enjoy the broadest view possible of the great lava field. As we reach the cloud deck, the sulfurous haze thickens and our view of the land fades away, ending our foray into the hot universe of Venus.

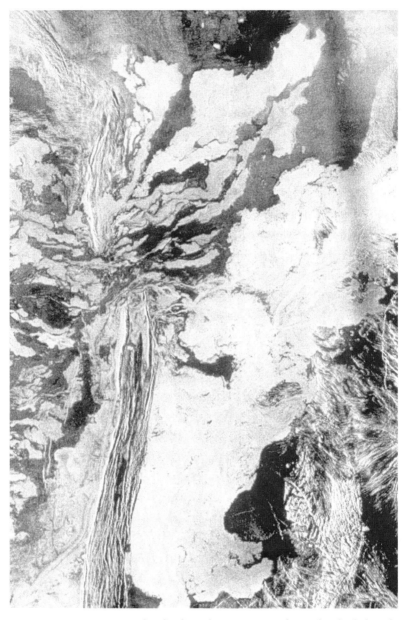

FIGURE 8.21 Lava breaks through a gap in a north–south ridge belt and fans out into the bright lava field of Mylitta Fluctus (right half of image). Credit: NASA/JPL-Caltech.

9 Volcanism on Io

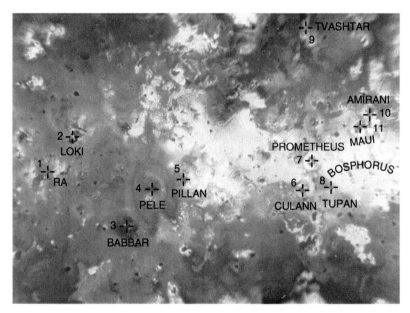

FIGURE 9.1 Map of Io. The major volcanoes discussed or illustrated in the text are indicated by numbers. (1) Ra Patera; (2) Loki; (3) Babbar; (4) Pele; (5) Pillan; (6) Culann; (7) Prometheus; (8) Tupan; (9) Tvashtar; (10) Amirani; (11) Maui. Credit: NASA/JPL-Caltech.

Ancient lava fields on the Moon, dormant lava shields on Mars, scores of volcanic features on Venus, but few hints of ongoing activity. As spacecraft after spacecraft explored the nearby planets, it appeared increasingly likely that the Earth was the only world in our Solar System that harbored eruptions today.

All this would change with a fortuitous discovery in March of 1979, when *Voyager 1* flew by Jupiter on its grand tour of the outer planets.

Few scientists expected to find active volcanism in the Jupiter system. The planet itself was out of the running. A giant ball of hydrogen and helium 143000 km in diameter, it lacked a solid surface. As for its harem of rocky satellites, four were comparable in size to our own Moon and, like our neighbor, had certainly run out of heat over the aeons. Most geologists expected their frozen surfaces to show only traces of ancient lava plains, punctured by impact craters.

Voyager 1 would provide the first glimpse of these icy worlds. On March 5, 1979, the American probe sped through the Jovian system on a trajectory that took it past the four largest moons: Callisto, Ganymede, Europa and innermost Io.

While the first two displayed ancient cratered landscapes as expected, the pair closest to Jupiter were most surprising. Europa had a bright surface of water ice, probably several kilometers thick, crisscrossed with ridges and grooves and showing very few impact craters. Arguably, some type of activity was smoothing out the crustal ice to erase the impact scars.

Io was even stranger. Because of its yellowish hue, noticeable through Earth-based telescopes, it was believed before the flight that Io was coated with colorful salts. But *Voyager*'s images surpassed the wildest expectations. Io's disk showed a patchwork of yellow and white, orange and red, laced with green, brown and black. It also featured circular spots, initially believed to be impact craters, although surprisingly dark.

But a bigger surprise was yet to come.

ERUPTIONS ON IO

As *Voyager 1* sped past Jupiter and its moons, mission controllers at the Jet Propulsion Laboratory checked its trajectory to make sure it was properly aimed at Saturn, its next destination. To check the spacecraft's bearing, navigator Linda Morabito called up a recent image of Io to measure the angle between Io's limb and a pair of reference stars in the background. She had to stretch the picture's contrast in order to pinpoint the faint stars above the moon's disk. As she did, a puzzling spot came into view, a crescent-shaped aura blossoming over Io's horizon. Could this be another moon, rising over the limb? A quick check showed that not to be the case. The only answer was that *Voyager 1* had witnessed a volcanic plume rising over the surface. The probe had recorded the first volcanic eruption ever observed on another world!

It was by no means a small eruption. The umbrella of ejecta rose some 270 km above the horizon and fell back over an area more than 1000 km wide – the size of Germany. In plan view, the plume was

FIGURE 9.2 Erupting volcano on Io, imaged in visible and ultraviolet light by *Voyager 1* on March 4, 1979. The outer part of the plume consists of finer sulfurous particles, visible only in ultraviolet. Credit: NASA/JPL-Caltech.

slightly asymmetrical – in the shape of a heart – and appeared to jet out from a dark spot on the surface. As more images were enhanced and analyzed, dark spokes were seen to radiate outward from the center – perhaps jets of coarser ejecta streaming away from the vent. In addition, two concentric rings surrounded the volcano – the inner one pale yellow and the outer one dark brown. Combined with the radial spokes, they created a peculiar dartboard motif.

The volcano and its plume were later named Pele, in honor of the fiery goddess of Hawaii. But Pele was not alone. On the same image where the first eruption was discovered, a second plume was spotted on the terminator, its particles illuminated by the grazing sunlight. Named Loki after a fiery Norse god, the eruption was traced to a fissure zone north of a dark caldera. It was resolved into two separate plumes: a larger one (about 200 km tall) at the western end of the fissure, and a smaller one (20 km tall) at the eastern end. Infrared data would later show Loki to be the most powerful volcano on Io.

The third erupting volcano detected on the *Voyager* pictures was baptized Prometheus. Its umbrella-shaped plume stood some 75 km tall and 250 km wide, with a dark central jet surrounded by a dartboard pattern of circular and radial markings, similar to those of Pele. Five more plumes were spotted on the processed imagery: Volund, Amirani, Maui, Masubi and Marduk. This brought the grand total to nine (Loki's double jet counting for two plumes).

The extraordinary heights reached by the plumes are due to the near vacuum reigning at the surface of Io and the moon's weak gravity field. Erupting gases expand dramatically as they leave the vent and reach velocities of 500 meters per second (perhaps even 1000 m/s at Pele), carrying particles to great height. As for the nature of the propelling gas, *Voyager*'s spectrometer showed Loki's plume to contain essentially sulfur dioxide (SO_2). The scattering properties of the Loki plume also gave away the size of the particles: several micrometers in the central jet and only a few nanometers (millionths of a millimeter) in the surrounding halo.

THE POWER OF TIDES

Io was expected to be geologically dead on account of its small size. Most of its initial heat should have bled off to space billions of years ago, as was the case on our own Moon (Io's diameter is 3680 km, versus 3470 km for the Moon).

The answer to this puzzle was actually formulated several months before the *Voyager* encounter and the discovery of its active volcanoes. In a prescient paper, physicist Stanton Peale calculated that the tidal pull exerted by Jupiter and by the other Galilean satellites on Io would wreak havoc inside its closest moon.

The process works as follows. Io spins on its axis in the same time that it circles Jupiter, thus keeping the same side permanently turned towards the planet. Moreover, Io's orbit is nearly a perfect circle. In such a stable configuration, Io should not feel any tidal stress from the giant planet and no active volcanism would be expected.

But an outside perturbation is forced upon Io by its sister moon

FIGURE 9.3 Io imaged by *Voyager 1* against the backdrop of Jupiter. Credit: NASA/JPL-Caltech.

Europa. Whereas Io circles Jupiter in 42 hours and 27 minutes, Europa orbits the planet in 85 hours – essentially double the period. As a result, Io overtakes Europa every other lap and feels a slight tug from its companion as it does. This causes it to depart ever so slightly from a circular orbit, with dire consequences. Jupiter's gravitational field is so strong that Io's slight yo-yo motion pumps up the tidal stress in its interior layers. The moon's surface might even bulge up and down by as much as 100 meters during each cycle. Dissipated as heat, this energy is what drives Io's outstanding volcanic activity.

This most original source of heat can be quantified. Calculations show that if Io is taken to be a homogeneous sphere, the tidal heating should reach 10^{12} watts – about ten times the heat produced inside our own Moon. Io's heat output would be even higher if it were a stratified body with a core, mantle and crust, as might be expected. Not only would this configuration amplify the tidal effect – generating closer to 10^{13} watts – but it would also focus it at the boundaries between layers (core–mantle and mantle–crust), where volcanism traditionally takes its source.

Hence, the volcanic plumes observed by *Voyager*'s camera were not a fluke, but the expression of a tremendous output of heat. The magnitude of this heat was confirmed by another instrument onboard the spacecraft: the IRIS infrared spectrometer.

PERSISTENT HOT SPOTS

Prior to the *Voyager 1* flyby, astronomers had already noticed that Io had a strange infrared behavior. On average, its surface temperature was slightly higher than what was expected at such great distance from the Sun, albeit it averaged little more than $-150\,°C$ (125 K) in the daytime and $-200\,°C$ (75 K) at night. Even more puzzling were temporary infrared "brightenings" that indicated temperature surges up to $+300\,°C$ (575 K) somewhere on the moon's surface.

Up close, *Voyager 1* confirmed that Io emitted an unusual heat flow, consistent with the tidal-heating model and the discovery of the volcanic plumes. The IRIS spectrometer detected nine "hot spots," several of which coincided with erupting plumes. Calculations showed that these areas had temperatures in excess of +100°C (375 K), with the hottest spot, centered on the volcano Pele, reaching close to +400°C (675 K).

Voyager 1 thus provided a surprising snapshot of a fiery world, as it sped past Io in March of 1979. But was this activity permanent, or just an exceptional crisis that happened to coincide with the probe's flyby?

Scientists were given the opportunity to tackle the question, because the *Voyager* mission involved two spacecraft. *Voyager 2* arrived at Jupiter in July of 1979, four months after its twin. Initially, *Voyager 2* was not scheduled to observe Io, but in view of the exciting *Voyager 1* discoveries, the second spacecraft was reprogrammed to turn its instruments toward Io during the flyby.

Despite a greater cruising distance, *Voyager 2* was able to locate seven out of the nine plumes observed by *Voyager 1*. The eighth plume was out of view of the camera, and the ninth – and not the least – was missing altogether. Pele, the largest eruption on Io during *Voyager 1*, had apparently shut down. Moreover, the pattern of ejecta around its vent was now fully circular, rather than heart-shaped as it appeared in the *Voyager 1* imagery. Sometime between March and July, Pele had "cleared its throat," blasting out whatever blocked its vent and ending its eruption in full symmetry.

Another volcano that evolved significantly between flybys was Loki, which showed different ground markings in July under its double plume. Loki's eruption was even seen to flare up as *Voyager 2* neared Jupiter, reaching heights of about 300 km before it settled back to around 150 km during the flyby. This time the eastern vent was the most active, spraying a much larger plume than the western vent.

No new plumes were detected on the *Voyager 2* imagery that were not already present in the *Voyager 1* data. But there were two new circular markings on the ground, which betrayed short-lived eruptions that flared up and died down in the four-month period between flybys. Apparently of the Pele type, since they were

FIGURE 9.4 The dark, horseshoe-shaped caldera of Loki dominates this view of Io obtained by *Voyager 1*. Just north of the caldera, a black bar with light-colored fans at each end is an erupting fissure with a double plume. Loki is the most active volcanic complex on Io. Credit: NASA/JPL-Caltech, processed by USGS Flagstaff.

short-lived and exceptionally large (the ejecta rings reached a diameter of 1400 km), these new volcanoes were baptized Aten and Surt.

From the data of both flybys, Loki turned out to be the largest and most powerful volcano on Io, albeit quite variable in its behavior. It occasionally radiates over 10^{12} watts of heat – more than all the volcanoes of Earth put together, and close to a quarter of Io's total output. In addition to the nine major hot spots, careful processing of the data brought out a dozen smaller features, much less energetic but still dwarfing anything known on Earth. Each of the "minor" hot spots of Io radiates ten to one hundred times more heat than the Hawaii or Yellowstone hot spot on Earth.

The *Voyager* spacecraft surveyed only one third of Io's surface. But extrapolating the heat flow to the entire moon (combined with additional data from Earth-based telescopes) suggests a global heat output of 10^{14} watts, about twice the heat output of the Earth, although Io's area is twelve times smaller. Unquestionably, Io is the "hottest" volcano world in the Solar System!

CALDERAS ON IO

Matching the infrared and the visible data showed that most hot spots on Io coincide with dark markings on the surface. Circular to irregular in outline, many of these spots are depressions that resemble calderas on Earth, although on a much larger scale. They are named paterae – a term assigned to such flat features on Venus and Mars.

The *Voyager* count lists 200 dark spots over 20 km in diameter. Their average size is 40 km, a few are over 100 km, and the largest one boasts a diameter of 200 km. In comparison, the largest caldera on Earth is Yellowstone, which is 77 km in diameter, and few are larger than 20 km. Large terrestrial calderas are formed by the collapse of magma chambers after voluminous eruptions of ash (these are associated with gas-rich, siliceous volcanoes).

Calderas also form on shield volcanoes, when magma withdraws and causes the summit to collapse. On Earth, "shield calderas" average only 6 km in diameter but on Mars – where lava conduits and magma chambers are much larger – shield calderas average 50 km in size, the record belonging to Arsia Mons that boasts a 100-km-wide summit depression. Io's calderas also appear to be collapse features. Many have scalloped edges and steep walls, typical of massive down drops. Shadow measurements indicate depths of a couple of kilometers in many cases: the record belongs to Chaac Patera with a 2.7 km drop from rim to bottom. But unlike calderas on other planets,

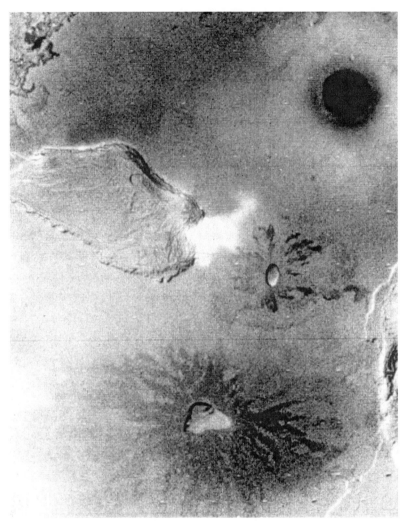

FIGURE 9.5 This *Voyager 1* image shows a dark circular caldera (top), a bright patch of SO_2 frost at the base of a cliff (center) with a volcanic shield to the right of it, and the oblong caldera of Maasaw Patera surrounded by dark lava flows (bottom). Credit: NASA/JPL-Caltech.

most calderas on Io lack an edifice and are flush with the plains. Only one in ten is perched on any noticeable relief, with radial lava flows pointing down slope.

Another generalization is that close to half of the paterae on Io lie at the foot of plateaus and steep blocks of crust that seem to have been tilted on edge. This association hints that tectonics play a major role in the formation of paterae. Rifting of the crust creates mountain blocks and opens up pathways for the magma to rise to the surface along the fault zones.

Most paterae on Io are very dark, allegedly because their flat bottoms are covered with fresh glassy lava. In fact, their intense heat flow indicates that they contain active lava lakes. The lava chills on the outside to form an outer crust, but the underlying heat still bleeds out each time the magma pushes through cracks and spills onto the surface. In the four months separating the *Voyager 1* and *2* flybys, for example, Loki's caldera underwent a color change: the southern half became much darker, as if it had received a new coating of fresh lava. The calderas of Surt and Aten also showed much darker floors in the *Voyager 2* imagery.

THE COLOR OF SULFUR

Many paterae on Io are surrounded by dark or bright lava flows that show convoluted margins and "fingers" extending out from the flow front – a pattern reminiscent of fluid pahoehoe lava flows on Earth. Many flow fields are close to 100 km in length, notably around the Prometheus caldera and on the gentle shield of Ra Patera (although at Ra, these flows were later obscured by more recent deposits). The record belongs to the Amirani flow field that stretches north to south over 400 km – comparable in size to provinces of flood basalt on Earth. Infrared readings show that the Amirani field is piping hot in places, which makes it the largest active flow field in the Solar System.

The Ionian lava flows have remarkable colors, in keeping with the moon's variegated aspect as a whole. Some are dark brown, others bluish-black, and others reddish to yellow. The flows of Ra Patera, for example, were seen to start out dark brown at the vent, turn red downstream, and fan out into orange lobes at their distal ends. Such a broad palette is reminiscent of the broad color range of elementary sulfur. Early on, this led to the hypothesis that Ionian flows were not molten silicates like basalt, but molten sulfur.

Sulfur flows are a rare occurrence on Earth, although sulfurous gases are common in volcanic settings. They rank third in abundance after water vapor and carbon dioxide. Because of its low melting and boiling points (112 °C and 445 °C respectively, about 300 K and 720 K), heated sulfur will tend to separate from silicate magmas and enter the gas phase. When the temperature drops, sulfur precipitates in fissures and vents to form yellow crystals and other colorful salts.

Occasionally, a moderate surge of heat will remelt sulfur crystals and generate a small flow of liquid sulfur. Such tongues of sulfur are seldom more than a few tens of meters long, but they exhibit features similar to those of basaltic flows: channels with levees, lobes branching away from the channel axis, and even miniature tunnels.

FIGURE 9.6 The volcanic plumes of Io are driven by sulfurous gases, expanding forcefully in the vacuum and low gravity of Io. Credit: artwork by William K. Hartmann.

FIGURE 9.7 On Earth, sulfur separates out of the magma and
precipitates in fracture zones. This fumarole sits in the crater of
Vulcano (Sicily). Sulfur changes color and behavior as its temperature
drops. The dark vent, a foot wide, shows both white sulfates and brown
molten sulfur. Orange sulfur lines the cavity and yellow sulfur coats
the periphery. Credit: photo by the author.

Sulfur flows, yellow to brown in color, have been observed
inside the dormant crater of Vulcano in Sicily. On Hawaii's Mauna
Loa, fumarole deposits of sulfur were melted by an eruption in 1950
to create a yellow flow 27 m long, 14 m wide and less than a meter
thick. At Lastarria volcano in the Andes, as well, there is a 350-m-
long sulfur flow that starts off greenish-yellow at its source and turns
grayish downstream. The largest sulfur flow used to lie on Japan's
Siretoko-Iosan volcano. Squeezed out during the 1936 eruption, it
started off chocolate-brown as it rushed down slope to form a tongue
1.4 km long and 20 m wide. Its color turned yellow–green as it solidified and cooled. Today, the flow has been mined and little of it
remains.

The color of sulfur is primarily a question of temperature. At its melting point – around 112 °C to 120 °C (385 to 393 K) depending on its crystal structure – sulfur is yellow. With increasing temperature, the melt turns orange at 130 °C (403 K), red–orange above 160 °C (433 K), and brownish-black over 220 °C (493 K). Inversely, hot sulfur might start off chocolate-brown and turn red, orange and yellow as it cools. To complicate the picture, it takes trace amounts of impurities, such as carbon, potassium or sodium to change the color of sulfur to greens, tans and grays.

Molten sulfur also displays a strange physical behavior as a function of temperature. Most molten substances like silicate magma are fluid at high temperatures and become more viscous as they cool and approach their solidification point. Sulfur behaves differently. As a hot chocolate-brown melt, it first behaves "normally" and becomes more viscous as it cools down to about 200 °C (475 K). But upon further cooling, as the molten sulfur turns orange, the trend suddenly reverses. Viscosity plummets and the melt becomes ten thousand times more fluid than when it was hotter and red! This anomaly only lasts over a short range of temperatures – between 200 °C and 150 °C (475 K and 425 K) – before the viscosity once again resumes a "normal" trend. As sulfur enters its yellow phase – below 150 °C (425 K) – its viscosity rises with falling temperature, until it finally solidifies around 115 °C (390 K).

The orange phase is truly treacherous. Imagine the surprise of a volcanologist on Io, watching an erupting sulfur flow. As it pours down the slope, the chocolate-colored flow cools and slows its progression, inviting the unsuspecting geologist to move in for a sample. But then, as the lava keeps cooling and turns orange, it suddenly speeds up into a gushing river and sends our volcanologist running for safety!

The picture of Io painted by the *Voyager* probes put sulfur in the starring role. Besides the colorful lava flows of Ra Patera and other volcanoes, there were spots of white "snow" scattered across the moon's surface that were fall-out crystals of frozen sulfur dioxide. Many such spots were located at the foot of scarps and cliffs, as if springs of liquid sulfur dioxide circulated underground, occasionally broke to the surface and sprayed a flurry of crystals across the landscape as it sub-limated. And weren't the erupting plumes of Prometheus, Loki and many other volcanoes precisely made of sulfur dioxide?

Other surface markings underscore the role of sulfur on Io. The red fall-out rings around Pele, Surt and Aten are best explained by jets of pure, unoxidized sulfur. Chemists point out that short chains of sulfur atoms (S_3 and S_4) are distinctly red. Due to the harsh radiation that bombards Io, these chains link together to form rings of eight atoms (S_8), which are yellow. Hence, red spots mark the location of recent fall-outs and yellow ones the older "riper" deposits.

THE HEAT RISES

In the wake of the *Voyager* flybys, sulfur appeared to many as being the principal agent of volcanism on the Io. Moreover, infrared data indicated temperatures under 400°C (700 K), consistent with molten sulfur, but harder to reconcile with basaltic magma, which is typi-cally hotter than 1200°C (1500 K).

Models were devised that involved mostly sulfur. Ponds of molten sulfur churned inside calderas, circulated through fractures and tubes, and fed the surface lava flows. To keep the pools hot, perhaps higher-temperature silicate magma erupted underground, on the floor of the sulfur reservoirs, but remained discrete and rarely emerged at the surface.

Critics pointed out that, despite plumes and colorful streaks at the surface, sulfur might play only a minor role in Io's volcanism. The plumes were indeed sulfur-based, and so was the red fall-out that sur-rounded calderas. But lava flows could still be molten silicates. It took little sulfur to react with cooling basalt and "colorize" its crust. As for the temperatures deduced from the infrared data, they were aver-aged over large areas. At the *Voyager* flyby distance, the resolution of the spectrometer was low. Each data point covered hundreds of kilo-meters squared, the area of large cooling lava fields. A piping-hot erupting vent, only hundreds of meters in size, was not expected to show up in the signal.

These burning questions would have to wait until another probe was sent to Jupiter, to observe its moons in greater detail. An ambitious spacecraft, named *Galileo*, was scheduled to be launched by the Space Shuttle, reach Jupiter, and fire a retro-rocket to lock itself into a Jovian orbit. The probe would then spend several years cruising from moon to moon, collecting close-up imagery and infrared data.

The tragic accident of the Space Shuttle *Challenger*, in 1986, postponed the program and imposed a longer trajectory that delayed the arrival of *Galileo* at Jupiter until December of 1995. In the meantime, however, advances in Earth-based astronomy filled the gap and brought in a host of surprises.

Although Io is a tiny dot in even the largest of telescopes, there are ways to monitor its volcanic activity. Io's infrared spectrum combines heat from reflected sunlight and heat from eruptions, but the signal can be analyzed to size up the latter. Moreover, when Io disap-

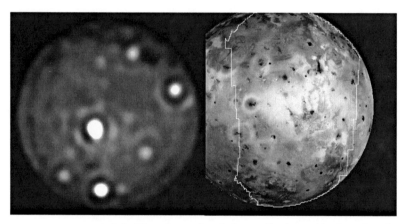

FIGURE 9.8 Left: The western hemisphere of Io, imaged in the infrared by the Keck telescope in Hawaii, using adaptive optics. The bright hot spot near the center is Loki. Right: The eastern hemisphere of Io seen in the infrared by the *Galileo* spacecraft (NIMS). Prometheus is the dark spot at center left. North and south, other hot spots create a "ring of fire" around a relatively cool area named Bosphorus Regio. Credit: (left) W. M. Keck Observatory/University California Berkeley; (right) NASA/JPL-Caltech, produced by the University of Arizona.

pears behind Jupiter (as seen from Earth), the infrared signal is progressively cut off, as the tiny disk glides behind the planet. The changing signal gives a reading of the heat distribution across the moon's surface. Spikes in the data can be correlated, longitude- and latitude-wise, with known volcanoes on the ground.

A first surprise came in 1988, when Io's infrared signal showed a spike at short wavelengths, indicating that somewhere on the surface, lava was exposed at temperatures in excess of 600°C (900K). This could not possibly be sulfur, which vaporizes at 440°C (715K). It was the proof that molten silicates flowed on the surface of Io.

Even more spectacular was the brief and intense outburst of energy from Io in January of 1990, which was traced to the Loki caldera. The signal implied temperatures up to 1200°C (1500K). It was later proposed that Loki's caldera harbored a lake of molten silicates, episodically turning over and exposing fresh magma in a burst of heat.

When *Galileo* finally reached Jupiter, in December of 1995, the goose was cooked. Io clearly bore the mark of high-temperature, basalt-like volcanism.

GALILEO AT JUPITER: IO REVISITED

Sixteen years had elapsed between the *Voyager* flybys of 1979 and the arrival of the *Galileo* orbiter in 1995. As it braked to enter an elliptical orbit around Jupiter, the spacecraft flew very close to Io, but problems with the onboard antenna and tape recorder prevented close-up observations of the fiery moon. The schedule then called for close encounters of the other Jovian moons: Europa, Ganymede and Callisto.

Io was out of bounds, because its tight orbit around Jupiter bathed it in the planet's intense magnetic field. Cruising through the radiation belt would be so hazardous for the probe's delicate electronics that close flybys of Io were pushed back to the end of the *Galileo* mission, towards 1999, when a failure would be less dramatic since the probe's mission would be nearly completed by that time.

During its first ten orbits around Jupiter (1996–97), *Galileo* never flew any closer to Io than 300 000 km – approximately the distance that separates the Earth from the Moon. But with its powerful long lens, *Galileo* was able to monitor Io's activity in almost as much detail as the closer *Voyager* flybys and over a much longer interval of time.

One priority was the search for plumes. The twin *Voyagers* had spotted nine erupting jets during their brief passes in 1979. Expectations were high that *Galileo* would see many more. Surprisingly, this was not the case. Only five of the nine *Voyager* plumes were still active, a quarter of a century later: Prometheus, Amirani, Marduk, Masubi (not spotted until 1998), and Pele (hardest to detect).

Volund, Maui and the two Loki plumes were conspicuously lacking. To compensate for the loss, six new plumes were spotted – in a five-year interval – over volcanoes Ra Patera, Culann and Pillan, and

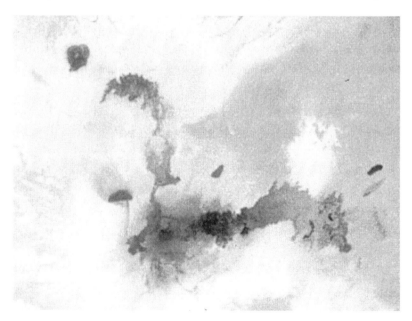

FIGURE 9.9 The lava fields of Maui (towards top) and Amirani (towards right), total 250 km in length. Bright halos of sulfur dioxide frost encircle the active areas. North is to the right. Image by *Galileo*. Credit: NASA/JPL-Caltech, produced by the University of Arizona.

over the lava fields of Zamama, Kanahekili and Acala. Most plumes were 50 to 100 km high, with Acala setting the record at nearly 300 kilometers.

Further insight into this spectacular phenomenon came from the surprising behavior of the Prometheus plume. New images by *Galileo* showed that the plume had shifted 80 km to the west, relative to its location in the *Voyager* pictures, 17 years prior. In the interim, connecting the old and new locations, an active lava field had spread across the cold sulfurous plains. White streaks of vaporizing SO_2 were later seen to highlight the flow fronts, as the hot lava crept over the frozen ground. One explanation for the plume's behavior would be that buried reservoirs of liquid SO_2 were being tapped at the flow front, vaporized and blown skyward, somewhere on the western edge of the Prometheus lava field.

In order to solve the mystery of the shifting plume, Prometheus was slated for the close-up passes over Io in 1999. Spectacular images of the flow front were collected, but they still failed to show the source of the 100-km-tall plume. To this day, the search for the elusive vent continues.

Besides Prometheus, the *Galileo* spacecraft spotted other areas that had changed their appearance during the 17 years between missions. There were new rings of bright-red material – sulfur-rich fall-out – which pointed to recent eruptions. Other spots were a darker shade of red, a "ripening" of the sulfur-rich deposits that attested to eruptions several years old. Some spots had even faded back into the background, such as the eruption rings of Surt and Aten, which *Voyager 2* had pictured in 1979. They had disappeared by 1995.

Galileo would now be watching Io for at least two years, and scientists hedged their bets. Catching the onset of a new, major eruption would be the icing on the cake.

THE PILLAN ERUPTION

Volcanologists got their wish granted. On the sixth orbit around Jupiter, in February of 1997, the *Galileo* probe picked up an infrared hot spot – among many others – over the Pillan caldera, 300 km east of Pele. Two orbits and three months later, in May of 1997, the *Galileo* probe imaged Io again and picked up a bright infrared flare. A major eruption was underway, with hot lava already spread over 70 square kilometers – the size of a major city. During the next observation of Io, in July, the lava field had grown to 400 km². Within it, an area of about 10 km² was piping hot: the infrared readings suggested temperatures above 600 °C (900 K).

This was clearly not sulfur, but rather a high-temperature silicate eruption. Thus, the model of basalt-like volcanism received a brilliant confirmation. As Io slipped into Jupiter's shadow, *Galileo*'s camera and infrared spectrometer scanned Pillan to obtain a tighter bracket on the lava temperature. In this eclipse situation, ideal for studying the glow of the eruption, the *Galileo* team derived peak temperatures above 1600 °C (1900 K), to which one might add an extra 200 to 300 K in order to pin down the true temperature of the magma. Indeed, the radiating body at the surface of Io had probably crusted over, at least in places, and had to be cooler than the initial melt.

Hence, this was an even hotter silicate than normal basalt, which erupts at temperatures of "only" 1200 °C (1500 K). One possibility was that the magma erupting at Pillan was superheated basalt, which rose so fast from the mantle that the drop in pressure caused its temperature to shoot up. Another possibility was that it was metal-rich *ultramafic* lava, created by extensive melting of a very hot mantle (with close to 40% of the source area turning into melt). Such metallic lava, named komatiite, was a common occurrence on Earth, billions of years ago, when our mantle was much hotter. Komatiites are full of iron and magnesium and are extremely fluid. Lavas at Pillan can be compared to the komatiites of South Africa, which probably erupted at temperatures of 1600 °C and contain 30% of magnesium oxide.

There was also a plume of ash rising 120 km above Pillan, spotted by both *Galileo* and Earth-based telescopes. The plume was certainly driven by sulfur dioxide gas, with some elemental sulfur

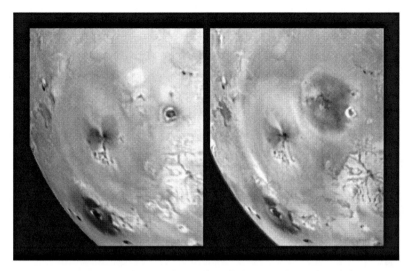

FIGURE 9.10 Pele is the heart-shaped plume at the center of these matching images, taken by *Galileo* in April 1997 (left) and September 1997 (right). In the latter, a new dark spot has blossomed in the upper right corner: the Pillan eruption. The dark mantle of pyroclasts covers an area the size of Arizona (400 km wide). North is at top. Credit: NASA/JPL-Caltech, produced by PIRL and the University of Arizona.

mixed in. What characterized it most was the amount of volcanic ash that it carried. In fact, the Pillan eruption stands out on the disk of Io mainly because of the quantity of dark fall-out. Centered on the plume, this "black eye" surrounding Pillan grew 400 km in diameter – the size of Arizona – in less than two months. The deposit probably represents five cubic kilometers of ash – five times the quantity expelled by Mount Saint Helens during its 1980 eruption.

The ash at Pillan bolstered the theory that the magma was ultramafic. A spectral analysis of the light reflected by the ash showed a dip in the near infrared, at a wavelength of 0.9 micrometer. This was the characteristic signature of magnesium-rich *orthopyroxene*, a mineral that thrives in ultramafic lava.

The Pillan eruption could then be pictured as a surge of ultramafic magma along a rift zone north of the caldera. Tens of kilometers in length, the fissure is marked by a streak of red sulfur particles.

Fountains of incandescent magma must have surged thousands of meters above the rift, in the near vacuum of Io. The plume of gas, mixed with glass and mineral shards, rose to an altitude of 150 km.

The curtain of fire fed a rapidly advancing lava flow. Two tongues of hot material headed southeast towards the caldera. Calculations hint that the eruption rate was greater than 2000 cubic meters of lava per second and probably close to the record eruption rate observed on Earth (8000 m³/s at Laki, Iceland, in 1783). Infrared observations showed two hot spots: one at the fissure zone, where the lava came out of the ground, and the other 50 km downstream of the source, where the flow cascaded over the caldera cliff.

Six months later, in March of 1998 when *Galileo* next observed Pillan, the floor of the caldera was covered with fresh lava. The eruption then died out. Later observations showed the hot spot to be fading. Pillan's activity had lasted one full year and treated volcanologists to the first ultramafic eruption ever witnessed.

GLOBAL IO

Besides Pillan, ultramafic eruptions appear to be commonplace on Io. Over its five years of active duty, *Galileo*'s infrared spectrometer detected a dozen hot spots with temperatures in excess of 1300 °C (1600 K). Besides these exceptional outbreaks, over twenty lava lakes and lava fields glow red-hot. Including "warm" areas, the number of hot spots detected on Io reaches a total of 166, with a higher concentration of the brightest ones around the equator than at the poles. Plume events also show a preference for low latitudes: 14 of the 15 plumes detected by *Voyager* and *Galileo* are located within 30° of the equator.

There also appears to be a bunching of hot spots in longitude, both around the sub-Jovian point (26 hot spots between 300°W and 360°W) and the anti-Jovian point (notably Prometheus and Culann). The sub-Jovian and anti-Jovian concentration of hot spots is readily explained by the distribution of tidal stresses – both compressional and extensional – that are exerted by Jupiter. Rupturing of the crust

FIGURE 9.11 Io is a patchwork of volcanic landscapes. Center left is the umbrella-shaped plume of Pele. Superposed on it is the dark blotch of the Pillan 1998 eruption. Other areas are covered with a light-hued frost of sulfur dioxide (upper left). Credit: NASA/JPL-Caltech, produced by PIRL and the University of Arizona.

occurs more frequently there, opening up pathways for magma to reach the surface.

Two competing models attempt to explain the creation and motion of magma inside Io. The first proposes that the tidal stresses melt the upper mantle – a 100 km to 200 km-thick asthenosphere just below the crust. Small convection cells within this "mushy" layer would distribute volcanic centers fairly randomly across Io. The second model favors melting of the deeper mantle. Larger, wider-spaced volcanic outlets would result, that would show a slight preference for polar regions. The apparent concentration of volcanoes around the equator appears to contradict the latter model and favor the shallow melting hypothesis, although a combination of both models is likely.

There is also some indication that Io's shallow mantle might be in a "super-molten" state – hotter and more liquid than the asthenosphere inside the Earth. Our asthenosphere is more solid than liquid: only 5% to 10% of its volume is melt. This proportion of melting yields a basic basaltic magma, relatively rich in alkali metals. If this melt undergoes further episodes of cooling and remelting (the process of differentiation), the resulting batches of magma grow increasingly richer in silica and alkali, yielding a whole suite of lighter-colored, lower density lavas.

On Io, volcanism has operated for so long that one would expect such a siliceous, low density "scum" to have collected at the surface over time. This is clearly not the case. On the contrary, most lavas on Io are believed to be basalts or komatiites, which are undifferentiated (low in silica). One solution to this paradox is that Io harbors a "magma ocean" in its upper mantle: a zone where over 20% of the rock is in a molten state. This super-hot layer is believed to be 200 km thick.

Hot komatiite erupts from this vast magma reservoir but has no time at the surface to evolve into more siliceous offspring. Fast buried by more recent flows, the layers of komatiite are pushed down to the base of the crust and re-enter the magma ocean, where they are mixed back into the batter. Or so goes the model.

Future spacecraft will want to land on Io and analyze the lava on the spot. Seismometers on several landers could set up a network and listen to the vibrations of the moon, delineate crust and mantle, and estimate the percentage of melt in the magma ocean. *Galileo* has raised as many questions as it has answered. Short of a landing, the probe ended its extraordinary mission by achieving six grazing flybys of Io, in 1999, 2000 and 2001.

THE I-24 FLYBY: PELE'S LAVA LAKE REVEALED

The first "barn-storming" pass of *Galileo* over Io took place on October 11, 1999. The probe flew over the southern tropics of Io, down to an altitude of 6000 km. As expected, the *Galileo* spacecraft took a severe beating from Jupiter's magnetic field as it closed in. Radiation damaged the electronics of the NIMS spectrometer, greatly reducing the number of wavelengths it could sample. The SSI camera also suffered from a glitch in its data compression software, which resulted in garbled imagery. Scientists worked hard to decipher the problem and unscramble the precious data.

Galileo's trajectory took it first over Loki and Pele, on the night side of Io.

Over Loki, the NIMS spectrometer scanned the central part of the caldera and detected cool temperatures, revealing that the lava flows were several months old, already capped by a thick insulating crust. The PPR radiometer, on the other hand, detected a hotter area in the southwestern part of the caldera, hinting that a new eruption was underway.

Over Pele, the SSI camera was turned on to look for any night-time glow that would betray the presence of a lava lake, churning inside the caldera. The results were startling. A dotted red line curved across the pitch-black image over a distance of 10 kilometers. It was less than 100 meters wide and its temperature read at least 700 °C (1000 K). At first, the dotted pattern suggested a collapsed lava tube with hot lava showing through a string of "windows." But the red line had a curved shape, apparently fitting the edge of the caldera. Thus it is believed to be the tumultuous margin of the lava lake, where the dark insulating crust jostles and breaks against the caldera wall, exposing red-hot magma. Such glowing shorelines are observed in lava lakes on Earth (see Figs. 9.12 and 9.13).

As *Galileo* flew over the terminator separating night from day, the nearby Pillan volcano came into view. The fresh lava field, emplaced during the spectacular 1997 eruption, showed a complex, rough surface in the grazing sunlight. Shadows along the flow front

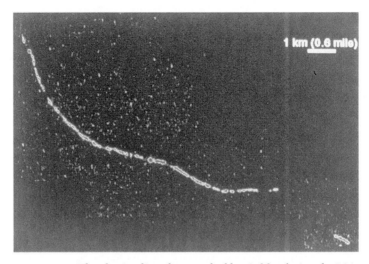

FIGURE 9.12 The glowing line photographed by *Galileo* during the I-24 nocturnal flyby over Pele is believed to be the glowing shoreline of a lava lake. The lava lake itself is crusted over and emits no light. Credit: NASA/JPL-Caltech, produced by the University of Arizona.

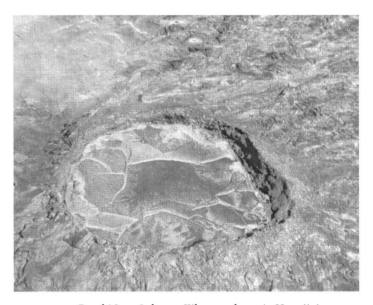

FIGURE 9.13 Pauahi Lava Lake, on Kilauea volcano in Hawaii, is a scaled-down version of Pele's churning caldera. The cooled crust of the lava lake tears apart along the shoreline, exposing hot magma. Credit: Hawaii National Park Service.

indicated that the lava sheet was approximately 10 meters thick. Small pits and domes were spotted across the lava field that reminded geologists of pseudocraters on Earth – rootless cones that were not true volcanoes but explosion "blisters." As they spread over volatile-rich terrain, lava flows are blown open as the trapped fluids vaporise and violently expand. On Earth, pseudocraters are caused by water flashing to steam. They are found on lava flows spreading over marshes and swampy coastlines. On Io, in the absence of water, such landforms are probably caused by the explosive vaporization of sulfur dioxide ice.

Before speeding away from the fiery moon, *Galileo* was able to observe volcanoes Zamama and Prometheus. Zamama was pictured showing a central shield, with radiating lava flows, and a long fissure zone – source of the most recent lava field. The lava field was covered with reddish fall-out from the central vent.

Scanned in high resolution for the first time, Prometheus revealed the intimate structure of its active flow field, dotted with bright breakouts of glowing lava. It was during this flyby that the NIMS spectrometer identified the hottest areas, both at the eastern and at the western ends of the field. They mark the vent area and the leading edge of the advancing flow front. The middle part of the field is cooler, because most of the lava flows underground in tunnels. The imagery failed to show the source of Prometheus' active plume, but only the southern half of the flow field had been imaged.

THE I-25 FLYBY: TVASHTAR'S CURTAIN OF FIRE

The *Galileo* team had little time to ponder before the spacecraft swung back for a second, even closer look at Io. On November 26, 1999, the probe flew over the south pole.

As it came in for its daring pass, down to a mere 300 km above the surface, *Galileo*'s electronics were again hit hard by Jupiter's radiation belt. This time, the spacecraft went into "safe mode" and automatically shut down its operations. Engineers rushed to reboot the system and *Galileo* came back on line half an hour after its closest approach. Most of the high-resolution imagery had been lost, but the probe secured beautiful color frames. Culann Patera, in particular, showed an elongate caldera, fanning lava flows, a long sinuous channel, and a red plume deposit. A mysterious green hue tinted the floor of the caldera and much of the flow field. It might result from the chemical interaction of the iron-rich lava with the sulfurous fall-out, creating greenish minerals – perhaps a form of pyrite.

Color imagery of Emakong caldera also showed a palette of stunning colors and convoluted lava flows and channels, reviving speculations that some flows on Io might consist of molten sulfur. The NIMS spectrometer confirmed that temperatures were so low at Emakong that cold patches of SO_2 frost survived amidst the flow field.

But the biggest surprise came from one of the last targets of the flyby: a panoramic scan of a chain of large calderas near the north pole, named Tvashtar Catena. An extremely bright streak tore across the central caldera, saturating the camera sensor. Careful analysis of the anomaly showed that the camera had caught a spectacular eruption on the caldera floor: a curtain of lava rising 1000 to 1500 meters above the surface, along a fissure 25 km in length!

Catching such a spectacular eruption in the act, during a short and spatially restricted flyby, was highly improbable, but the *Galileo* team had struck gold. The NIMS spectrometer scanned the eruption in the western half of the fissure zone and derived a minimum temperature of 700 °C (1060 K). Backing up the spacecraft, Earth-based telescopes analyzed the infrared brightening of Io due to the eruption and derived temperatures up to 1600 °C (1900 K), hinting again that the lava was magnesium-rich komatiite. In order to follow the progress of the eruption, the *Galileo* team targeted Tvashtar Catena for a follow-up observation three months later, during the probe's third pass over Io.

THE I-27 FLYBY: CALDERAS AND LAVA FLOWS
The last flyby of the series took place on February 22, 2000, along a trajectory that encompassed most of the targets from the two

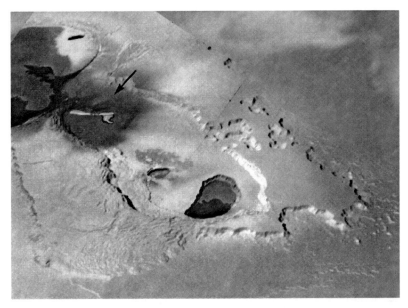

FIGURE 9.14 In the Tvashtar Catena row of craters, *Galileo* caught an eruption in the act (left) during its skimming pass of November 1999. Credit: NASA/JPL-Caltech, produced by the University of Arizona.

FIGURE 9.15 A curtain of fire jets out of a rift zone in Hawaii. The magma is carried upward by expanding gas. Credit: Hawaii National Park Service.

previous runs. In this way, observations could be updated and eruptions followed up over a four-month period.

Miraculously, the I-27 encounter was spared the electronic glitches and computer resets that had plagued the previous flybys. A new night-time pass over Pele found the same glowing line that marked the edge of the lava lake. Its geometry was unchanged. On the other hand, the NIMS spectrometer detected another breakout of hot magma in the southeastern section of the caldera, where temperatures of about 1500°C (1760K) were measured, again much higher than the melting point of basalt (1250°C). Hence, Pele joined Pillan and Tvashtar in the select list of "very hot" volcanoes that were suspected to erupt komatiite lava.

Chaac Patera received special attention on the I-27 pass. One of the largest calderas on Io, Chaac shows many similarities with Hawaiian calderas on Earth, except for its much larger size. It also appears that the lava rises through rim fractures along the edge of the caldera, rather than randomly on its floor – which illustrates the importance of extensional tectonics on Io. Pits, lava mounds and "bathtub rings" testify to a complex history of inflation and deflation of lava flows, filling the caldera, venting out their gas and draining back into their vents. Pockets of bright material hint that SO_2-rich liquid might also pond in the cooler areas of the basin.

The lava fields of Prometheus and Amirani were also scrutinized during the I-27 flyby. The Prometheus images were compared to those collected during I-24, allowing geologists to map out new lava tongues that had squeezed to the surface over the four-month interval. A total of 60 km² of new lava was measured, which translates into an average eruption rate of 0.5 km²/day, ten times the resurfacing rate of the current Kilauea eruption in Hawaii. As for the elusive SO_2 plume, again no single vent source could be identified. Apparently, Prometheus-type plumes formed from the blending together of multiple scattered jets of vaporized SO_2 along the advancing lava flow, rather than from one large, detectable vent.

As for Amirani – the largest active flow field in the Solar

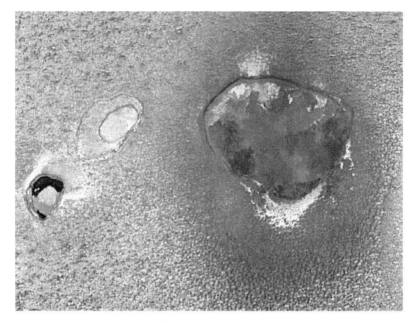

FIGURE 9.16 *Galileo* flew over Camaxtli Patera on its I-27 flyby (February 22, 2000). The caldera floor is covered with dark recent lava flows. Bright patches mark the site of escaping sulfurous gas. Two smaller pits are at left. Credit: NASA/JPL-Caltech, produced by the University of Arizona.

System – comparing the I-24 and I-27 imagery revealed $620\,km^2$ of new lava, a $5\,km^2/day$ rate that is 100 times the resurfacing rate at Kilauea volcano. Although impressive, this lava output is still well below the maximum discharge rates observed during the largest eruptions on Earth. Discharge rates at Amirani do not exceed $500\ m^3/s$, whereas the 1789 Laki eruption in Iceland reached a maximum rate of $8000\ m^3/s$.

From one flyby to the next, another area of change was the Tvashtar chain of calderas, where *Galileo* had caught the spectacular lava fountain, three months prior. The February 2000 image showed that fountaining had ceased in the central caldera, but that a new blinding streak of incandescent lava ran along the margin of the westernmost caldera. The 60-km-long flow front appeared to be fed from two different vents. The hot lava was surrounded by a dark deposit, probably ash expelled during a previous phase of fountaining activity.

THE 2001 FLYBYS

After an 18-month truce, during which it flew by two other moons of Jupiter (Ganymede and Callisto), *Galileo* returned for a final series of buzzing passes over Io – its swan song before the termination of its seven-year mission.

In December 2000, as it prepared for its final run, the probe detected a tenuous plume over Tvashtar – the site of the fire curtain a year prior – that rose 385 km above the surface. The I-31 flyby called for a low altitude passage over Tvashtar and there was some concern that the spacecraft might be damaged by the plume particles.

When it buzzed over Tvashtar Catena on August 6, 2001, the plume had vanished, but an even larger plume shot up from a new

FIGURE 9.17 During its final pass over Io in October 2001, *Galileo* took a last image of Tvashtar Catena. The 2000 eruption has ceased in the central caldera, and spokes of bright deposits from the spent plume radiate into the plains. Credit: NASA/JPL-Caltech, produced by the University of Arizona.

site, 600 km to the south, betrayed by the presence of a whopping new hot spot. The sulfurous jet soared 500 km above the surface – a record height – and the spacecraft flew through the edge of its broad umbrella, recording hits from the scattered particles. Based on their mass, the impactors appeared to be snowflakes of sulfur dioxide, containing 15 to 20 molecules clumped together. Unscathed, *Galileo* went on to record temperatures over Tvashtar proper, creating an infrared map of the cooling lava field in the westernmost caldera. Finally, as it receded from Io, the spacecraft imaged a new plume deposit – a 1000-km-wide ring around Dazhbog Patera – and completed an infrared scan of the giant Amirani lava field.

After broadcasting its data back to Earth and completing one more loop around Jupiter, the spacecraft swooped in for an even closer pass over Io, on October 16 (I-32 flyby). During the approach, it had a last night-time look at Pele and Loki and their putative lava lakes. The camera and infrared spectrometer showed areas of high temperature (1100 °C or 1400 K) where vigorous currents broke up the crust and exposed hot magma. Loki's horseshoe caldera also showed breakouts of warm lava along its western edge.

As *Galileo* reached the sunlit side of Io, it pointed its instruments at two colorful volcanoes: Emakong and Tupan. Emakong is a candidate location for low-temperature sulfur volcanism on Io. The high-resolution imagery focused on a lava channel that showed an intricate mix of bright and dark colors, strengthening the sulfur flow hypothesis. In the following chapter, we visit Emakong on one of our Ionian field trips (see pages 326–330).

As for Tupan, named after a Brazilian thunder god, its 75-km-wide caldera revealed tall cliffs (900 m high, according to shadows) and a large "island" in the center of the depression, surrounded by dark lava. A red coating covers most of the island and caldera floor, probably sulfur sprayed out of erupting vents. Green patches could indicate areas where the sulfur reacted with hot lava, and yellow patches along the caldera wall might be sulfur deposits that oozed out of the cliffs.

FAREWELL TO IO

As *Galileo* receded from Jupiter's fiery moon, it completed an infrared tally of hot spots with its infrared spectrometer, nailing over a dozen additional heat sources and bringing the grand total to 166.

On January 17, 2002, *Galileo* swooped in for its final barnstorming pass over Io, but Jupiter's radiation belt once again battered the onboard electronics and incapacitated the spacecraft. No data were collected and *Galileo* silently flew by Io for the last time, its seven-year mission coming to an end. On September 21, 2003, in order to eliminate any chance that the uncontrolled probe might one day hit Europa (and contaminate this potentially life-bearing moon), *Galileo* was sent plunging to its death into the thick atmosphere of Jupiter.

The spacecraft observation of Io is over for the while being, but monitoring of its volcanic activity continues with Earth-based telescopes. Developments in adaptive optics that cancel the distortions due to the Earth's turbulent atmosphere enable ground-based telescopes to image Io in the infrared with a resolution of 100 km per pixel.

Astronomers at the Keck Observatory in Hawaii detected three powerful hot spots on Io in February 2001, which match up with the known locations of Amirani, Tvashtar and Surt volcanoes.

The Surt eruption in particular was spectacular. The volcano had not erupted since the *Voyager* flybys of 1979, and its 2001 outburst gave off a heat flux of 7.4×10^{13} watts – arguably the largest eruption ever witnessed on Io. Large fire fountains are implied, with calculated temperatures up to 1200 °C (1475 K). Since then, another gigantic eruption has been traced to Tupan caldera.

Besides this telescopic vigil, robotic spacecraft will ultimately return to Io, the volcanic champion of the Solar System. And one day, as we develop in the next chapter, astronauts will behold the wonders of Io with their own eyes.

A tour of Ionian volcanoes

Pele

Latitude:	19° S
Longitude:	257° W
Size of caldera:	24 km × 10 km
Diameter of plume:	>1000 km
Altitude of plume:	280 to 460 km
Age:	currently active

FIGURE 10.1 Pele displays a heart-shaped set of rings around a dark center. Lower left lies the dark spot of Babbar Patera. Credit: NASA/JPL-Caltech.

Pele was the first volcanic eruption to be detected on Io. In March of 1979, *Voyager 1* imaged an umbrella-shaped plume rising 270 km over the surface. Named after the fiery goddess of Hawaii, the volcanic center consists of a dark elongated caldera, in a rift zone that hugs the edge of Danube Planum, a triangular-shaped plateau. The rift and plateau are the expression of tilted blocks of crust and underscore the dominant role of tectonics, which opens pathways for deep-seated magma to rise to the surface. No lava flows have been spotted outside the caldera, although we lack detailed imagery of the area. The visible activity seems confined to the caldera, which is believed to be a crusted-over lava lake.

Nighttime flybys over the caldera by the *Galileo* spacecraft have identified a sinuous glowing line in the southwestern corner, where the crust breaks against the shore of the lava lake and exposes the red-hot lava beneath it. Hot magma must also break out in other areas on the caldera floor, to account for the high heat flow emanating from Pele – an estimated 230 gigawatts of power. These powerful

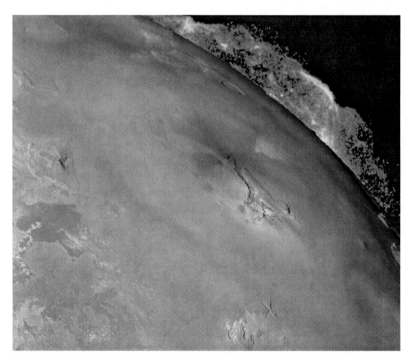

FIGURE 10.2 Pele erupting. The erupting rift is visible in the center, surrounded by rings of sulfurous deposits from the plume. The plume itself is visible above the horizon, rising to an altitude of 300 km. Credit: NASA/JPL-Caltech, produced by the U. S. Geological Survey.

heat outlets have not yet been located, but temperatures of at least 1400 °C (1700 K) are suspected there. Elsewhere, the lava crust is several meters thick and has cooled to –100 °C (175 K) in the chilly vacuum of Io.

The eruption plume that rises over Pele probably blows out of one of the many undiscovered fissures on the lava lake. The driving gas is sulfur dioxide (SO_2) with a few percent of monoxide (SO) and elemental sulfur. Calculations show that the gas must reach velocities close to one kilometer per second as it jets out of the vent, entraining fine particles to altitudes of 250 km and more. In October, 1999, the Hubble Space Telescope measured a plume height of 350 kilometers. *Galileo* caught a glimpse of the plume in 1996, when it reached a record height of 460 km.

Pele's plume is difficult to observe, due to the small size of the entrained particles (around ten micrometers in diameter). Particles must be larger during major blow-outs, as indicated by the ejecta rings surrounding the caldera.

Two dark fan-shaped deposits stretch 200 km east and west of the caldera, drawing an hourglass pattern across the plains. The spectral signature of orthopyroxene – a magnesium-rich silicate – has been detected in the deposits. Farther out, past a few yellow patches, a ring of red material stretches out to an overall diameter of 1400 km. The red hue is probably due to particles of sulfur linked in chains of three or four atoms (S_3 and S_4) – the red variety of sulfur. This more viscous material probably needs greater thrust to be expelled from the vent and consequently flies out to farther distances.

In summary, Pele is essentially a lava lake of molten silicates, contained within a rift zone caldera, and associated with a high-energy plume driven by magmatic gas. In the infrared, Pele is a major hot spot, radiating on the order of 200 to 300 gigawatts of power. In order to radiate so much heat, the upward flow of magma into the lava lake must be on the order of 300 cubic meters per second. Pele stands out among plume-producing volcanoes on Io by being the only one so far that has been caught in the act of producing a large red ring. This particularity, as well as its supersonic gases, sets Pele's plume aside with respect to most other plumes on Io.

Like all features on Io, Pele is extremely young. It is impossible to say how long ago it became established as a major volcanic center. Since its discovery in 1979, Pele appears to erupt non-stop, as attested by its constant heat flow and permanent plume. We start our Ionian tour with a visit to this landmark, extraordinarily active volcano.

Itinerary

We use a westward trajectory to approach Pele for a landing. This takes us over wide plains of sulfur dioxide frost – glistening with a pale lavender hue. Upon final approach, we fly over the large dark spot of the Pillan eruption, 300 km east of Pele.

Pillan's eruption was imaged in the summer of 1997. Silicate lava flows surround the caldera, as well as a 400-km-wide blanket of dark particles around its caldera. We take advantage of the low altitude fly-over to collect imagery of the fresh lava field. But its details are already erased by a blanket of red sulfur particles, which Pele sprays in a bright vermilion ring, up to 600 km from its vent. It is this tenuous plume which we now penetrate, with instruments deployed to sample its gas and dust.

As we close in on our target, we drift a little south to get a good view of the Danube plateau: a 200-km-wide block of uplifted crust, torn apart by a Y-shaped rift. We loop around the plateau to enter the rift from the south and fly northward along its floor, looking for landslides or layering that might reveal the nature of the Ionian crust. Reaching the mouth of the valley, we hook around a spur to catch a grand view of the Pele caldera from above, shaped like a smoking pipe. In the foreground stands a main basin, 20 km in diameter. Behind it, a "stem" extends to the west, along a narrow rift zone that hugs the northern face of the Danube plateau. The dark lava lake occupies the southern half of the basin. We aim for a landing spot on the older, inactive part of the lake, but still within reach of the action.

As we slip into our spacesuits, we are reminded of the hazards

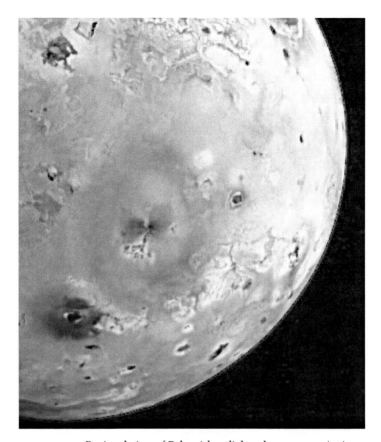

FIGURE 10.3 Regional view of Pele with radial spokes, concentric rings and the dark spot of Babbar Patera to the southwest. Credit: NASA/JPL-Caltech, produced by PIRL and the University of Arizona.

awaiting us. The low gravity – one sixth that of the Earth – will feel familiar to moonwalkers, but there are many differences with the Moon. The lighting is marginal at best. At Jupiter's distance from the Sun, the illumination is 30 times weaker than on the Earth and Moon. The vivid colors are still breathtaking. At our feet, the dark lava shows streaks of sulfurous salts: yellow, orange and even pale green.

Much more of a concern than the lighting is the invisible solar wind – energetic protons and electrons that are channeled down to Io's surface by Jupiter's intense magnetic field. Without protection,

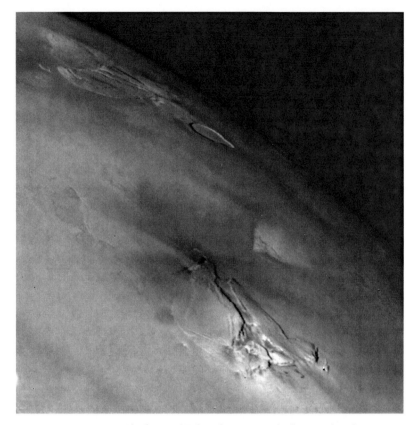

FIGURE 10.4 The heart of Pele volcano is a rifted triangular plateau, named Danube Planum. Along its northern margin, a dark spot marks the location of the hot caldera. Fuzzy spokes radiate from the erupting vent. Credit: NASA/JPL-Caltech.

we would amass a lethal dose of radiation in less than a minute. Our specially designed spacesuits and pressurized rovers cut down the exposure considerably, but our hours are counted and we must make the most of our short EVA (Extra Vehicular Activity), before retreating to the safety of our spacecraft.

We ride towards the active zone of the caldera to the south, over the glassy lava flows, and stop several times along the way to collect samples. The thick cool lava is covered with decades of dusty fall-out – specks of sulfur and volcanic glass that rained out of the erupting plume. As we look up at the sky, we make out swirls and bands of gas and dust that remind us of northern lights on Earth – a ghostlike pattern of shock waves in the stream of particles that rise, fall and mingle above us. Where does the plume take its source? It appears denser above the rift zone that extends the caldera to the west. We have no hope of reaching that far, because a formidable obstacle will stop our progression.

As we move towards the darker and more recent caldera floor, covered with a thinner layer of ash, we begin to make out a fiery glow illuminating the plume from below. This is the southwestern edge of the lava lake, where the churning magma manages to break through the rafts of solid crust and evacuate much heat and gas. The lava crust that we tread upon is probably no more than ten meters thick, and covers molten magma. It is just a few years old and has had time to cool to a chilly $-100\,°C$ (175 K). But at times, our sensors pick up puffs of heat sifting through cracks – a reminder of our tenuous position, tiptoeing on the thin scum of a lava lake.

The glow in front of us has materialized into a blinding, jagged line of red magma, tearing the surface. As we approach the incandescent lava, erupting from a fissure on the caldera floor, our infrared radiometer registers a minimum temperature of $1000\,°C$ (1300 K). We stop at a reasonable distance from the 100-meter-wide wall of fire and watch fountains of magma shoot up, propelled by unpredictable pulses of gas. The spectacle reminds us of the curtains of fire and lapping lava lakes of Hawaii, but there is an eerie feeling of slow motion as the clumps of lava fall back at only a fraction of the speed that they do on Earth. With this comes the realization that we are truly on another planet, witnessing our first eruption away from home.

Loki

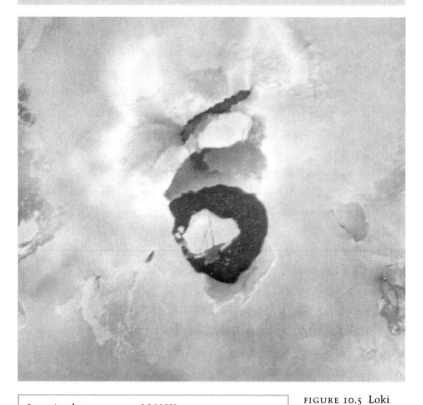

Longitude:	309° W
Diameter of caldera:	200 km
Height of plumes:	150 to 400 km (Loki-west)
	20 to 150 km (Loki-east)
Heat output:	$1-2 \times 10^{13}$ W
Age:	currently active

FIGURE 10.5 Loki displays a central caldera and an elongate fissure zone to the north, both darkened by recent lava. Credit: NASA/ JPL-Caltech, processing by the U. S. Geological Survey.

Named after a mischievous Norse god – condemned to everlasting pain and twitching – Loki is the largest caldera and the most energetic hot spot on Io. Infrared measurements show that the volcanic center radiates over 10^{13} watts of heat – about a quarter of the moon's entire heat flow and more than all of the Earth's hot spots put together.

This tremendous energy emanates from a dark-floored caldera 200 km in width, with a central "island" of bright material that splits the basin into a horseshoe. The dark floor of the caldera averages a warm +30 °C (300 K), as compared to the freezing −150 °C (120 K) of the surrounding plains.

Some areas of the Loki caldera are even hotter. Infrared surveys show that temperatures reach at least 300 °C (600 K) around the margins of the caldera floor and along the coast of the central island. This thermal pattern is best explained by a lava lake, which brews under a thin, solidified crust. Rafted about, the crust breaks along the walls of the caldera and against the shores of the island, exposing hot magma there.

Major eruptions occur often. Every year or two, infrared observatories report that Loki has a "brightening," when its heat output increases up to ten-fold in less than a month, then remains at a high level for another few months before returning to normal. A sheet of hot lava that spreads across a section of the caldera floor can best explain this pattern.

Some events are even more spectacular. Known as "outbursts," they flare up in a couple of hours, doubling the heat output *of the entire moon*, before fading out in less than a week. Outbursts are best explained by a dramatic, global overturn of the lava lake, the crust foundering to be replaced by hot magma over a wide area.

Despite the tremendous activity at Loki, there are presently no obvious fall-out rings of glass or sulfur particles. The only plumes in the vicinity were imaged by the *Voyager* probes in 1979 and can be traced to a linear fissure, 300 km northeast of the caldera. Named Loki-west (19°N, 305°W) and Loki-east (17°N, 301°W), the two plumes rose to heights of 150 km at the time. In ultraviolet imagery, the finer particles of Loki-west were even seen to reach 400 km. These ephemeral plumes, although they bear the name Loki, might belong to an independent volcanic center.

There are no fresh lava fields around Loki. All activity seems confined to the giant caldera. Because of its size and energy, Loki is

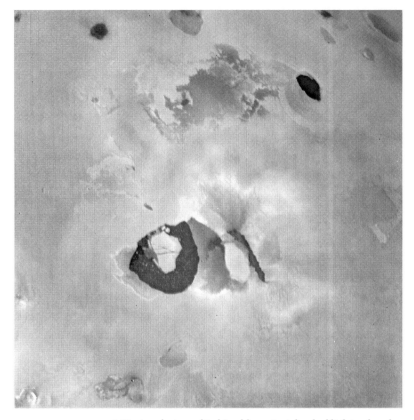

FIGURE 10.6 Regional view of Loki caldera. An island of light-colored crust, gouged by fissures, is surrounded by a horseshoe-shaped pool of dark lava. Other calderas are visible at the top of the image. Credit: NASA/JPL-Caltech.

considered to be a long-lasting volcanic center on Io. As the most powerful volcano in the Solar System, it is a mandatory destination for the intrepid interplanetary voyager.

Itinerary

The descent towards Loki is spectacular, with the huge globe of Jupiter looming in our portholes as we swing the spacecraft towards our landing spot. We approach from the north, flying over the long rift that hosted active plumes, back in the *Voyager* days. Discrete color streaks on the ground mark the location of ancient ash falls.

FIGURE 10.7 The Loki plume rises over the horizon in this flyby shot taken by *Voyager 1*. Credit: NASA/JPL-Caltech, processing by the U. S. Geological Survey.

Coming up on Loki, with the small pale Sun low on the horizon, we are surprised by the shiny quicksilver hue of the caldera floor, so different from the dark tones of conventional imagery. This is due to the grazing illumination bouncing off the glassy, mirror-like lava. We fly low over the lava crust, picking out cracks and fissures coated with colorful sulfur deposits, watching ghostly whiffs of gas jet out in the near vacuum of Io.

A blinding white plateau looms straight ahead: the central island of Loki. Why has this 50-km-wide platform survived in the center of the caldera? Is it firmly rooted in the crust, or does it float like a raft on the molten magma?

We fly over the white plateau, coated with a frosting of sulfur dioxide. Two parallel cracks, a couple of kilometers wide, cut across the island. We follow the fractures to the south, turning on cameras and spectrometers to peer at the bottom. This is a narrow rift flooded with fresh lava. Soon the end of the plateau is in sight and we brake our spacecraft for a soft landing on the edge of the cliff, overlooking the southern branch of the caldera floor. It would be suicide to land on the floor itself, since eruptions at Loki are frequent and unpredictable – the largest ones overturning big sections of the lava lake in cataclysmic "outbursts." Could we be so lucky as to witness one of these rare but spectacular eruptions from afar?

From the safety of the plateau summit we enjoy an exceptional view of the margin of the lava lake, where the cool glistening crust jostles against the shores of our "island." It breaks up in places to reveal bright orange magma beneath it.

Our short outdoor trek – in our radiation-resistant spacesuits – gives us an opportunity to pick up some cores of the sulfur dioxide frost and soil, and to search for any chunk of rock that would reveal to us the nature of the island crust. The best place to get an idea of its structure and composition is the wide crevice that we followed on our way in. Standing on its precipitous edge, we take imagery and spectral readings of the various layers that jut out of the facing wall. Sheets of unprocessed ultramafic lava, or evolved silica-rich "scum?"

We also take time to install the first seismometer of a regional geophysical array, in order to monitor earthquakes and, by analyzing the strength, speed and direction of the seismic waves, to generate a 3-D image of the Ionian crust. We drive a thermometer into the ground to measure the heat flow, expecting figures in excess of 5 watts per square meter. Once the geophysical set-up is completed, our time is up in the radiation-drenched environment. We retreat to the safety of our landing module, with the huge globe of Jupiter hanging still on the western horizon.

Prometheus

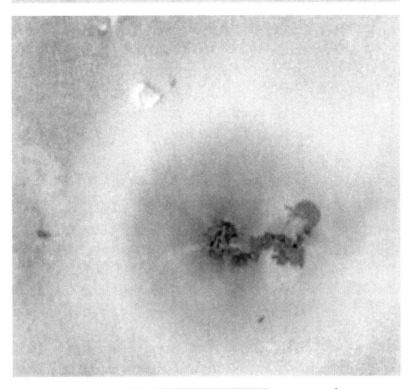

Latitude:	2° S
Longitude:	154° W
Size of caldera:	28 km × 14 km
Height of plume:	100 km
Heat output:	10^{11} watts
Age:	currently active

FIGURE 10.8
Prometheus volcano consists of a caldera shaped like a swan's head (right), and a hook-shaped lava field (darker, left). A plume and spokes of sulfurous gas emerge from the western end of the lava field. Credit: NASA/JPL-Caltech, produced by the University of Arizona.

Prometheus is the most active volcano on Io's anti-Jovian hemisphere – the side of Io turned away from Jupiter. The caldera and its lava field are associated with a bright plume that rises 100 km above the surface. Because the umbrella of particles has appeared on every image of the area so far – taken both by *Voyager* in 1979 and *Galileo* since 1996 – the Prometheus plume is known as the "Old Faithful" of Io.

Over the years, however, the location of the plume has shifted 80 km to the west, away from the caldera, and appears to follow the flow front of the advancing lava. This observation was key in establishing that the plume – and many others on Io – are created by the vaporization of snow fields of sulfur dioxide (SO_2), as the lava creeps over the frosted-over soil.

The lava emerges from a fissure zone 15 km south of the caldera, where minimal temperatures of 700 °C (1000 K) have been reported. The lava runs to the south, then to the west. The flows have lobate, crenulated margins, highlighted by streaks of SO_2 that stream away from the flow front. Breakouts of fresh lava appear on the edge, but also in the middle of the flow field, suggesting that the field is an "inflated compound flow," fed by lava tubes and puffing up vertically as well as spreading out horizontally. At the western distal end of the lava field, temperatures exceeding 300 °C (600 K) betray outbreaks along the lava front. The exact location of the plume within the lava field is not known and constitutes a major objective of our field trip. The 100-km-tall plume drops an inner ring of bright material in the plains – probably a powder of elemental sulfur – and an outer ring of sulfur dioxide, up to 300 km from the vent.

Prometheus is the site of a major hot spot. Its heat output reaches one hundred gigawatts (10^{11} watts). It is nearly one hundred times weaker than Loki (the undisputed record holder), but it is one of its bright runners-up on Io. It dominates the anti-Jovian hemisphere and appears to be part of a larger regional ensemble. An infrared survey of the area by the *Galileo* NIMS spectrometer shows a dozen smaller hot spots in the vicinity of Prometheus – including

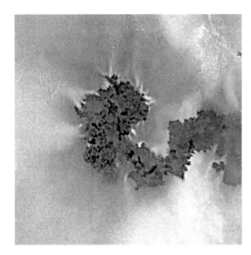

FIGURE 10.9 The Prometheus lava field grew 100 km westward between the *Voyager 1* flyby of 1979 and this image taken by *Galileo* in February 2000. Bright spokes of sulfurous gas jet out from the advancing flow front. Credit: NASA/JPL-Caltech, produced by the University of Arizona.

Culann to the south and Tupan to the southeast: the area appears to have twice the average concentration of hot spots observed elsewhere on Io.

Because of its active lava field and constantly erupting plume, Prometheus is perhaps the most exciting, erupting volcano to visit on Io.

Itinerary

We come down for a night landing at Prometheus, just before dawn, heading west over the equator. In our descent, we cross the "Ring of Fire" – a circular (perhaps random) line-up of hot spots that includes the Amirani lava field and the calderas of Tupan and Culann. We fly over the cold Bosphorus Regio that lies in the center of the ring and aim for the red glow of Prometheus that lights up the western horizon. To the south, we can also make out the red glows of Culann and Tupan, like two cities lit up in the distance.

From a distance, Prometheus looks like the head of a swan. Two lava flows diverge from the 20-km-wide caldera to form an open beak. On automatic pilot, with floodlights turned on for safety, we land in the neck of the swan – a patch of frost-covered plains that lies between the caldera and the lava field. The vent area that feeds the

FIGURE 10.10 The lava field (left) is a patchwork of different hues, pointing to many breakouts of lava. The plains to the right are covered with sulfur dioxide frost, vaporized by the advancing lava front and coating the ridges in the plains. Credit: NASA/JPL-Caltech, produced by the University of Arizona.

1 km (0.6 mile)

80-km-long lava flows is only a couple of kilometers south of our landing spot. Preparing for EVA, we can watch the bright glow of the vent light up the night sky through our southern window.

As dawn breaks, we step out onto the white sulfurous plains. Something is missing: the giant colorful globe of Jupiter no longer hangs over our head. This is our first outing on the anti-Jovian hemisphere, which never gets to see its tyrannical parent planet.

On our roving tour of Prometheus, our first stop is at the vent. From orbit, infrared spectrometers like NIMS measured minimal temperatures of 600 °C (900 K). On the ground, a member of our party climbs a hill of lava to get a clear view of the vent and aim a field spectrometer at the glowing magma, in order to get a direct reading of the temperature. It is expected to be close to 1500 °C (1800 K).

We sample what appears to be the oldest lava in the area, covered with particles of red sulfur, as well as a much fresher-looking flow. We are eager to find out if the lava is basalt or the higher-temperature, magnesium-rich komatiite. These first samples are too glassy and vesicular to harbor large crystals that would give us an answer. We will have to wait for the verdict of a chemical analysis, back in the lab.

After reboarding the pressurized rover, we drive out from the vent area onto the smooth frosty terrain that borders the lava field, and drive westward. Over the next 80 km, we ride along the convoluted flow front – a ten-meter-high wall of glistening rock, plastered in places with sulfurous fall-out. Lobes of smooth lava spill out in some places and mounds of rubble pile up in others.

Occasionally, we run across a fan of gouged soil, streaked with pebbles, that marks the blow-out site of a pocket of sulfur dioxide frost, overridden by the lava. We stop at such a site to sample precipitates, including sulfur. The ridged plains along the lava flow are likewise coated with a sugar-like spray of sulfur dioxide frost.

As we reach the extremity of the lava flow, we still have been unable to locate the source of the Prometheus plume, although the air looks bluer and denser towards the impenetrable inner reaches of the lava field. Somewhere in the midst of the buckled lava lies the elusive vent, draining the pressurized liquid sulfur dioxide out of the heated ground.

Rather than risk a trek across the flow field, we stay along the edge and look for fresh breakouts of lava, where toes of metallic melt creep onto the sulfurous surface and blow up like silent firecrackers, frost flashing to gas in the vacuum of Io. After watching scores of small explosions along the flow front, we terminate the EVA and return to the safety of the pressurized rover, with more questions than answers . . .

Ra Patera

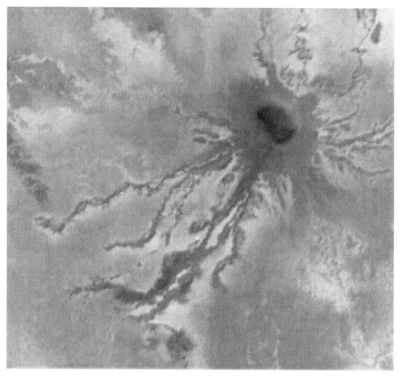

Longitude:	325° W
Diameter of shield:	450 km
Area of lava:	250 000 km²
Relief:	less than 1000 m
Diameter of caldera:	35 km
Height of plume:	75 km
Heat output:	undetected
Age:	currently active

FIGURE 10.11 Ra Patera and its digitate lava flows, viewed by *Voyager 1*. Credit: NASA/JPL-Caltech, processed by LPI and Paul M. Schenk.

Ra Patera is one of the most aesthetic and mysterious volcanoes on Io. On *Voyager* imagery it has the plan form of a shallow shield, centered on a 35-km-wide caldera. The distribution of lava flows is asymmetric. The longest ones stretch out westward, like the tentacles of a giant squid. The older lava flows that constitute the shield are wide and bright. They cover an area of 250000 km^2 – the size of the United Kingdom – but the lava shield rises less than 1000 m above the surrounding plains.

The superposed, dark and narrow flows are the most intriguing. They are up to 250 km long and less than 5 km wide, except in one location where two branches merge to create a 15-km-wide lava delta. Their reddish tint first convinced scientists that they were sulfur flows, rather than silicate lava. This model is no longer favored, since we now know that most active flows on Io are hotter than the boiling point of sulfur. On a local scale, however, it is still conceivable that small bright lobes, branching out of the main flows, are run-offs of molten sulfur. These run-offs are so numerous that they create a bright halo along the darker flows.

Although Ra Patera was quiet during the *Voyager* flybys, it awakened in 1994 when the Hubble Space Telescope spotted a brightening of the area, which also turned yellow. Upon arrival at Jupiter, the *Galileo* probe reported in 1996 that a new dark field of lava had indeed surrounded the caldera and flowed to the southeast. White deposits extended farther out – possibly fall-out from a plume. And a plume there was: spotted in June 1996, it rose to an altitude of 75 km. By the time of the next observation, in November, the plume had vanished or at least had fallen below the detection level of the camera.

The 1994–96 eruption of Ra Patera did not produce a detectable flash of heat. *Galileo*'s infrared spectrometer failed to detect a hot spot in the area. Either the magma breakout had shut down at the time of the observation (with the plume still going on strong), or else the magma was unusually "cool," reawakening the specter of low-temperature sulfur flows. To solve this mystery, a landing at Ra Patera will constitute an exciting milestone in our exploration of Io.

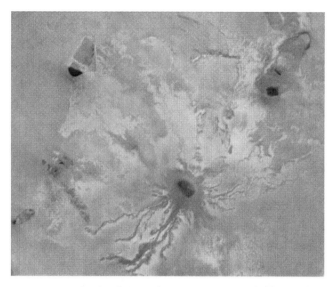

FIGURE 10.12 Regional view of Ra Patera, surrounded by patches of bright sulfurous frost. Credit: NASA/JPL-Caltech, produced by the U. S. Geological Survey.

Itinerary

Ra Patera lies in the sub-Jovian hemisphere of Io, a short distance from Loki (to the northeast) and Pele (to the southeast). The area has substantial relief, as we learn on our westward descent into the target area. Our radar altimeter indicates a crescent-shaped ridge 5 km above the plains, dropping to a low plateau that abuts Ra Patera's eastern flank. The 1994–96 lava flows emerge from the caldera and curl around the obstacle. We fly along the southern scarp of the flow and pick a landing spot on the edge of the dark lava field, upon smooth plains that shine bright yellow in the early morning light.

One of our first assignments, as we exit our module, is to sample the yellow soil. We drive a core tube into the ground, which is arguably the sulfur-rich fall-out of the plume spotted in 1996. Once the core is extracted and packaged, we board our rover for a westward traverse toward the narrow "tentacle" flows that were spotted on the *Voyager* imagery and sparked the sulfur-flow theory.

During the long ride out west, staring out the front window, we struggle to recognize landmarks on our map, realizing how fast the landscape changes under the dusty breath of Io's volcanoes. The most recent fall-out is close to a meter thick and coats the area with patches of yellow and orange. Nonetheless, we soon make out the subdued flow front of one of the "tentacles" – a long and narrow strip of lava.

We stop our rover at the foot of the ash-covered wall. According to our readings, we are close to the breakout point of a subsidiary flow. We exit in our bulky spacesuits for a brief walk to the flow front, hopping across the site like *Apollo* moonwalkers. We stare at the ash-covered wall facing us. Is the flow made of silicate stock like basalt? Is it a tongue of sulfur that oozed out of the ground when it was over-ridden by hot silicate? Or is it mainly sulfur that constitutes the entire lava field?

In order to secure representative samples, we might want to test a scaled-up version of a hair dryer, using pressurized nitrogen to dust off the outcrops. Blowing away the sulfurous fall-out, we work on a large boulder that rolled down from the flow front. Komatiite, basalt or solid sulfur? Place your bets.

FIGURE 10.13 A plume of sulfurous gas jets out over Io, while Jupiter hangs still in the sky. Credit: artwork by William K. Hartmann.

Emakong Patera

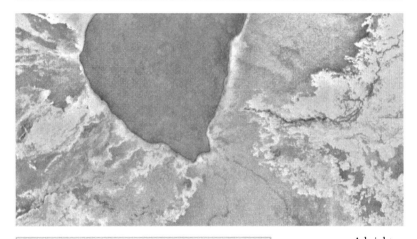

Latitude:	4° S
Longitude:	120° W
Diameter of caldera:	66 km
Depth:	230 m
Longest lava flow:	370 km
Heat output:	low
Age:	active

FIGURE 10.14 A bright lava field spills out of Emakong caldera towards the east. Credit: NASA/JPL-Caltech, produced by David Williams, Arizona State University.

Emakong Patera is a dark caldera surrounded by bright lava flows, on the anti-Jovian side of Io. The complex is located near the equator in the middle of the frost-covered region of Bosphorus Regio. Although surrounded by active volcanoes ("The Ring of Fire"), Bosphorus Regio itself shows little thermal activity. Only a modest hot spot was detected at Emakong caldera by the *Galileo* spacecraft as it flew by in 1999, and the temperatures inferred are less than 100 °C (400 K). There has been no change in the overall appearance of Emakong, and thus no major eruption, since the first images taken by the *Voyager* spacecraft in 1979. Emakong's caldera is heart shaped and averages 66 km in diameter. Shadow measurements indicate that the caldera walls are up to 230m high in places. Although the heat flow from the caldera is relatively low, there is a slight enhancement along the margins, as would be expected from a crusted-over lava lake that breaks its crust and bleeds off heat along the battered shoreline.

Emakong is surrounded by both bright and dark lava flows, most of which seem to have spilled out of the caldera during high stands of the lava lake. Whereas dark flows on Io are believed to be silicate lavas (such as basalt or komatiite), bright flows such as those around Emakong, yellow to white in colour, are much more enig-matic. The length of these lava flows (one extends for 370km north of the caldera) and their convoluted margins both point to low-viscosity lava. This low viscosity could be the result of a very hot sil-icate magma, just as it could be the mark of a much lower temperature but fluid sulfur melt.

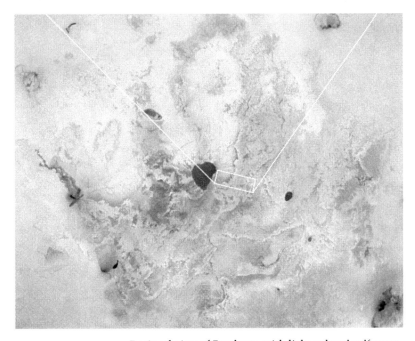

FIGURE 10.15 Regional view of Emakong, with light-colored sulfurous fall-out over the lava fields. Box shows area of close-up (Fig. 10.16). North is to the top. Credit: NASA/JPL-Caltech, produced by the University of Arizona.

One other puzzling detail is that *Galileo*'s infrared spectrometer detected the signature of cold sulfur dioxide frost on the caldera floor, in areas that also radiate heat. This overlap is astonishing, since sulfurous frost is not expected to survive too close to an active flow. One must assume a very patchy layout of the caldera floor, with discrete breakouts of lava spacing wide tracks of sulfurous frost.

Several sinuous channels spring from fissures in the ground near the rim of the caldera and feed an extensive field of bright lava flows. One channel in particular emerges 10 km down slope from the caldera's eastern rim, strikes east for 50 km, and then turns north. The dark channel is 105 km long and averages 500 m in width – a slightly scaled-down version of Hadley Rille on the Moon.

In close-up, the mottled appearance of the channel and of its lava field suggest that volcanism at Emakong mixes both silica-based and sulfur-based stock. An *in situ* investigation might help us sort out their respective contributions.

Itinerary

We fly over the frosty plains of Bosphorus Regio as we come in for a landing on the outer slope of Emakong Patera. On our final approach we skim across the caldera floor, looking for breakouts of dark lava amidst the bright patches of sulfurous frost. The black and white mottled motif, with shades of brown and pale lavender, looks like a giant quilt, keeping the underlying magma nice and warm.

As we reach the eastern margin of the caldera, fissures and lava breakouts grow more numerous, but we fail to identify the nature of the melt, laced with streaks of bright orange and chocolate brown, like an exotic ice cream sundae. The steep caldera wall looms ahead, looking very much like the cliff of Kilauea crater in Hawaii. We strain to make out details in Emakong's wall. Our first impression is that the wall face is too steep to be solid sulfur, which we would expect to fail and slump plastically. Aprons of brown and yellow debris might well be tongues of sulfur that oozed out of the cliffs, but otherwise the rigid fabric of the caldera wall appears to be made of solid rock.

FIGURE 10.16 The Emakong lava field, east of the caldera, is crossed by a dark sinuous channel. Strips of high-resolution imagery are superposed on a low-resolution background image. North is to the top. Credit: NASA/JPL-Caltech, produced by the University of Arizona.

Skimming over the top of the cliff, we find ourselves gliding down the gently sloping outer flank of Emakong – a field of dark grey lava, laced with hues of lilac and gold. Straight ahead we spot the dark course of the main lava channel, surrounded by a fan of bright flows.

The dark channel is over 500 m wide, which appears too large for a sulfur flow. As we fly down the lava course, we notice how it seems to cut through the bright lava flows, probably the mark of a very hot turbulent flow that dug through the substratum. Our control panel flashes briefly, signaling the landing site straight ahead. The channel opens up into a wider reach, bearing two streamlined islands in its center. We skim along the largest island – a full kilometer long and 200 m across – and land in its lee on the flat lava bed.

From the ground, the island shows a sharp streamlined face, cut by the hot fluid that swept down the course. This cliff face is an ideal location to study the history of the lava field and scan the different layers that the channel incised. Do we spot ash falls and sulfur flows between the thick layers of silicate lava?

We embark on a short drive down the channel to reach a breath-taking viewpoint: spanning the entire width of the channel, a roof of solid rock transforms the open course into a giant lava tube. The mouth of the tunnel is choked with slabs of foundered roof segments. Even in the low gravity of Io, these great arches of solidified lava come regularly crashing down, and we restrain ourselves from driving too deep into the tunnel.

Staying near the entrance, we drive up to a slab of collapsed roof to lop off a sample and check if it is made of metal-rich silicate, as expected. The floor of the tunnel might harbor a few surprises: veins of colorful sulfur, or perhaps even streaks of nickel or silver that have separated from the melt and collected into seams and pockets. What better way to end our tour of the Solar System than by collecting a few souvenirs – Ionian jewelry – for our friends back home?

Bibliography

BOOKS ON PLANETS AND VOLCANOES

Beatty, J. K., Petersen, C. and **Chaikin, A.**, eds. (1999). *The New Solar System*. Cambridge: Cambridge University Press.

Carr, M. H., ed. (1984). *Geology of the Terrestrial Planets*. SP-469. Washington, DC: NASA.

Cattermole, P. (1989). *Planetary Volcanism: a Study of Volcanic Activity in the Solar System*. Chichester: Ellis Horwood Ltd.

Cattermole, P. (1994). *Venus, The Geological Story*. Baltimore: Johns Hopkins University Press.

Francis, P. (1993). *Volcanoes: A Planetary Perspective*. Oxford: Clarendon Press.

Frankel, C. (1996). *Volcanoes of the Solar System*. Cambridge: Cambridge University Press.

Girault, F. *et al.* (1998). *Volcans Vus de l'Espace*. Paris: Nathan.

Greeley, R. and **Batson, R.** (2001). *The Compact NASA Atlas of the Solar System*. Cambridge: Cambridge University Press.

Hartmann, W. K. (2003). *A Traveler's Guide to Mars: The Mysterious Landscapes of the Red Planet*. New York: Workman Publishing.

Hodges, C. A. and **Moore, H. J.** (1994). *Atlas of Volcanic Landforms on Mars*. USGS Professional Paper 1534.

Lopes, R. (2004). *The Volcano Adventure Guide*. Cambridge: Cambridge University Press.

Macdonald, G. A. (1972). *Volcanoes*. Englewood, NJ: Prentice Hall.

Sheehan, W. (1996). *The Planet Mars: A History of Observation and Discovery*. Tucson: University of Arizona Press.

Sigurdsson, H. *et al.* (2000). *Encyclopedia of Volcanoes*. New York: Academic Press.

Simkin, T. and **Siebert, L.** (1994). *Volcanoes of the World*. Tucson: Geoscience Press.

Taylor, S. R. (1982). *Planetary Science: A Lunar Perspective*. Lunar and Planetary Institute.

Wilhelms, D. (1993). *To a Rocky Moon: A Geologist's History of Lunar Exploration*. Tucson: University of Arizona Press.

SCIENTIFIC LITERATURE

General

Wilson, L. and Head III, J. W. (1981). Ascent and eruption of basaltic magma on the Earth and Moon. *Journal of Geophysical Research*, **86**, 2971–3001.

Wilson, L. and Head III, J. W. (1983). Comparison of volcanic eruption processes on Earth, Moon, Mars, Io and Venus. *Nature*, **302**, 663–9.

Wood, C. A. (1984). Calderas: a planetary perspective. *Journal of Geophysical Research*, **89**, 8391–406.

Chapter 1

Clague, D. A. *et al.* (2000). Near-ridge seamount chains in the northeastern Pacific ocean. *Journal of Geophysical Research*, **106**, 16541–61.

Coffin, M. F. and Edlholm, O. (1993). Scratching the surface: estimating dimensions of large igneous provinces. *Geology*, **21**, 515–18.

Greeley, R. (1982). The Snake River Plains, Idaho: representative of a new category of volcanism. *Journal of Geophysical Research*, **87**, 2705–12.

Heikinian, R., Moore, J. G. and Bryan, W. B. (1976). Volcanic rocks and processes of the Mid-Atlantic Ridge rift valley. *Contrib. Mineral. Petrol.*, **58**, 83–110.

McCormick, P. *et al.* (1995). Atmospheric effects of the Mt. Pinatubo eruption. *Nature*, **373**, 399–402.

Rampino, M. R. and Caldeira, K. (1992). Episodes of terrestrial geological activity during the past 260 million years: a quantative approach. *Celestial Mechanics and Dynamical Astronomy*, **54**, 143–59.

Simkin, T. (1993). Terrestrial volcanism in space and time. *Annu. Rev. Earth Planet. Sci.*, **21**, 427–52.

Sleep, N. H. (1992). Hotspot volcanism and mantle plumes. *Annu. Rev. Earth Planet. Sci.*, **20**, 19–43.

Smith, D. K. and Cann, J. R. (1992). The role of seamount volcanism in crustal construction at the Mid-Atlantic Ridge. *Journal of Geophysical Research*, **97**, 1645–58.

Chapter 2

Harris, A. J. L. *et al.* (2000). Effusion rate trends at Etna and Krafla and their implications for eruptive mechanisms. *Journal of Volcanological and Geothermal Research*, **102**, 237–69.

Pinkerton, H. *et al.* (2000). Exotic lava flows. In *Environmental Effects on Volcanic Eruptions: From Deep Oceans to Deep Space*, eds. J. R. Zimbleman and T. K. P. Gregg. New York: Kluwer Academic/Plenum Publishers.

Chapter 3

Head III, J. W. (1976). Lunar volcanism in space and time. *Rev. Geophys. Space Phys.*, **14**, 265–300.

Schmitt, H. H. (1991). Evolution of the Moon: *Apollo* model. *American Mineralogist*, **76**, 773–84.

Wood, C. (1975). The Moon. *Scientific American*, **233**, 93–102.

Chapter 4

Greeley, R. (1971). Lava tubes and channels in the lunar Marius Hills. *The Moon*, **1(2)**, 237–52.

Head III, J. W. and **McCord, T. B.** (1978). Imbian-age highland volcanism on the Moon: the Gruithuisen and Mairan domes. *Science*, **199**, 1433–6.

Howard, K. A. and **Head III, J. W.** (1972). Regional geology of Hadley Rille. In *Apollo 15 Preliminary Science Report*. SP-289. Washington, DC: NASA.

Schmitt, H. H. and **Cernan, E. A.** (1973). A geological investigation of the Taurus-Littrow Valley. In *Apollo 17 Preliminary Science Report*. SP-293. Washington DC: NASA.

Spudis, P. D. and **Greeley, R.** (1987). The formation of Hadley Rille and implications for the geology of the *Apollo 15* region. *Lunar Planet. Sci.*, **XVIII**, 243–54.

Swann, G. A. *et al.* (1972). Preliminary geologic investigation of the *Apollo 15* landing site. In *Apollo 15 Preliminary Science Report*. SP-289. Washington DC: NASA.

Wagner, R. *et al.* (2002). Stratigraphic sequence and ages of volcanic units in the Gruithuisen region of the Moon. *Journal of Geophysical Research*, **107**, 5104.

Weitz, C. and **Head III, J. W.** (1999). Spectral properties of the Marius Hills volcanic complex and implications for the formation of lunar domes and cones. *Journal of Geophysical Research*, **104**, 18933–56.

Whitford-Stark, J. L. and **Head III, J.W.** (1980). Stratigraphy of Oceanus Procellarum Basalts: sources and styles of emplacement. *Journal of Geophysical Research*, **85**, 6579–609.

Chapter 5

Carr, M. H. (1976). The Volcanoes of Mars. *Scientific American*, **234**, 32–43.

McEwen, A. S. *et al.* (1999). Voluminous volcanism on early Mars revealed in Valles Marineris. *Nature*, **397**, 584–86.

McKay, D. S. *et al.* (1996). Search for life on Mars: possible relic biogenic activity in Martian meteorite ALH84001. *Science*, **273**, 974–30.

Rieder, R. *et al.* (1997). The chemical composition of Martian soils and rocks returned by the mobile alpha proton X-ray spectrometer: preliminary results from the X-ray mode. *Science*, **278**, 1771–4.

Wilson, L. *et al.* (2001). Evidence for episodicity in the magma supply to the large Tharsis volcanoes. *Journal of Geophysical Research*, **106**, 1423–33.

Wilson, L. and Mouginis-Mark, P. J. (1987). Volcanic input to the atmosphere from Alba Patera on Mars. *Nature*, **330**, 354–7.

Wyatt, M. B. and McSween, Jr., H. Y. (2002). Spectral evidence for weathered basalt as an alternative to andesite in the northern lowlands of Mars. *Nature*, **417**, 263–6.

Chapter 6

Crown, D. A. and Greeley R. (1993). Volcanic geology of Hadriaca Patera and the eastern Hellas Basin, Mars. *Icarus*, **100**, 1–25.

Morris, E. C. and Tanaka, K. L. (1994). *Geologic maps of the Olympus Mons region of Mars*. U.S. Geological Survey Miscellaneous Investigations Map I-2327, scale 1/2 000 000.

Mouginis-Mark, P. J. *et al.* (1982). Explosive volcanism on Hecates Tholus, Mars: investigation of eruption conditions. *Journal of Geophysical Research*, **87**, 9890–9904.

Mouginis-Mark, P. J. *et al.* (1988). Polygenic eruptions on Alba Patera, Mars. *Bulletin of Volcanology*, **50**, 361–79.

Watters, T. R. *et al.* (1987). Distribution of strain in the floor of the Olympus Mons caldera. In *MEVTV Workshop on the evolution of magma bodies on Mars*. LPI Technical Report, **90-04**, 293–4.

Chapter 7

Bullock, M. A. and Grinspoon, D. H. (March 1999). Global climate change on Venus. *Scientific American*, **280**, 50–7.

Crumpler, L. S. *et al.* (1996). Volcanoes and centers of volcanism on Venus. In *Venus II*, eds. S. W. Bougher *et al.* Tucson: University of Arizona Press.

Head III, J. W. *et al.* (1991). Venus volcanism: initial analysis from *Magellan* data. *Science*, **252**, 276–88.

Head III, J. W. and Wilson, L. (1992). Magma reservoirs and neutral buoyancy zones on Venus: implications for the formation and evolution of volcanic landforms. *Journal of Geophysical Research*, **97**, 3877–903.

Head III, J.W. *et al.* (1992). Venus volcanism: classifications of volcanic features and structures, associations and global distribution from *Magellan* data. *Journal of Geophysical Research*, **97**, 13153–98.

Koch, D. M. (1994). A spreading model for plumes on Venus. *Journal of Geophysical Research,* **99**, 2035–52.

Turcotte, D. L. (1993). An episodic hypothesis for Venusian tectonics. *Journal of Geophysical Research,* **98**, 17061–8.

Chapter 8

Keddie, S. T. and **Head III, J. W.** (1994). Sapas Mons, Venus: evolution of a large shield volcano. *Earth, Moon and Planets,* **65**, 129–90.

Magee Roberts, K. M. *et al.* (1992). Mylitta Fluctus, Venus: rift-related, centralized volcanism and the emplacement of large-volume flow units. *Journal of Geophysical Research,* **97**, 15991–16015.

Chapter 9

Crown, D. A. and **Greeley, R.** (1986). Sulphur and volcanism on Io. *Nature,* **322**, 593–4.

Keszthelyi, L. *et al.* (1999). Revisiting the hypothesis of a mushy global magma ocean on Io. *Icarus,* **141**, 415–9.

Keszthelyi, L. *et al.* (2001). Imaging of volcanic activity on Jupiter's moon Io by *Galileo* during the *Galileo* Europa mission and the *Galileo* Millenium mission. *Journal of Geophysical Research,* **106**, 33025–52.

Lopes-Gautier, R. M. C. *et al.* (1999). Hot spots on Io: global distribution and variations in activity. *Icarus,* **140**, 243–64.

Lopes, R. M. C. *et al.* (2001). Io in the near infrared: near-infrared mapping spectrometer (NIMS) results from the *Galileo* flybys in 1999 and 2000. *Journal of Geophysical Research,* **106**, 33053–78.

Lopes, R. M. C. *et al.* (in press). Lava lakes on Io: observations of Io's volcanic activity from *Galileo* NIMS during the 2001 flybys. *Icarus.*

McEwen, A. S. *et al.* (1998). Active volcanism on Io as seen by *Galileo* SSI. *Icarus,* **135**, 181–219.

McEwen, A. S. *et al.* (1998). High temperature silicate volcanism on Jupiter's moon Io. *Science,* **281**, 87–90.

McEwen, A. S. *et al.* (2000). Extreme volcanism on Jupiter's moon Io. In *Environmental Effects on Volcanic Eruptions: From Deep Oceans to Deep Space,* eds. J. R. Zimbleman and T. K. P. Gregg, New York: Kluwer Academic/ Plenum Publishers.

Peale, S. J. *et al.* (1979). Melting of Io by tidal dissipation. *Science,* **203**, 892–4.

Radebaugh, J. L. *et al.* (2001). Paterae on Io: a new type of volcanic caldera? *Journal of Geophysical Research,* **106**, 33005–20.

Spencer, J. R. *et al.* (1997). Volcanic resurfacing of Io: post-repair HST imaging. *Icarus*, **127**, 221–37.

Williams, D. A. *et al.* (2001). Evaluation of sulfur flow emplacement on Io from *Galileo* data and numerical modeling. *Journal of Geophysical Research*, **106**, 33161–74.

Chapter 10

Davies, A. *et al.* (2001). Thermal signature, eruption style and eruption evolution at Pele and Pillan on Io. *Journal of Geophysical Research*, **106**, 33079–105.

Kieffer, S. W. *et al.* (2000). Prometheus, the Wanderer. *Science*, **288**, 1204–8.

Schenk, P. *et al.* (1997). Geology and topography of Ra Patera, Io, in the *Voyager* era: prelude to eruption. *Geophysical Research Letters*, **24**, 2467–70.

Williams, D. A. *et al.* (2002). High-resolution views of Io's Emakong Patera: latest *Galileo* imaging results. In *Lunar and Planetary Science XXXIII*, Abstract #1339. Houston: Lunar and Planetary Institute.

WEBSITES

Volcano World (University of North Dakota)
  http://volcano.und.nodak.edu/vw.html
Hawaiian Volcano Observatory (Hawaiian volcanoes)
  http://hvo.wr.usgs.gov/
Italy's volcanoes (Boris Behncke, Catania)
  http://boris.vulcanoetna.com/
USGS/Cascades Volcano Observatory (all volcanoes)
  http://vulcan.wr.usgs.gov/Volcanoes/
Smithsonian Institution's Volcanoes of the World (data base)
  http://www.volcano.si.edu/gvp/world/
Planetary photojournal (imagery of all planets, JPL/NASA)
  http://photojournal.jpl.nasa.gov/index.html
Views of the Solar System (Calvin Hamilton)
  http://www.iki.rssi.ru/solar/eng/homepage.htm
Volcanology on Mars (NASA Ames)
  http://amesnews.arc.nasa.gov/erc/MarsVolc/
Mars volcano imagery, *Mars Global Surveyor* (Malin Space Science Systems)
  http://www.msss.com/mars_images/moc/MENUS/volcanic_list.html
Io volcanic imagery (*Galileo*, JPL/NASA)
  http://galileo.jpl.nasa.gov/images/io/ioplume.html

Glossary

aa lava

texture of a lava flow, in which the surface is composed of blocky chunks.

accretion

process by which planetary bodies collide and fuse together. It is one of the principal mechanisms that lead to the creation of planets.

alkali basalt

a form of basalt that is often rich in *olivine* crystals. It is characteristic of *hot spots* and has its source deep in the *mantle.*

alkaline suite

evolution of a *magma* that yields lava types increasingly rich in alkali metals (sodium and potassium). Typical of *hot spot* settings.

andesite

lava type, slightly more siliceous than basalt, that contains *plagioclase feldspar* and *pyroxene* crystals. On Earth, andesite is typical of *subduction* settings (named after the Andes volcanoes).

anorthosite

siliceous rock made principally of anorthite crystals – a calcium-rich feldspar.

arachnoid

class of volcanoes on Venus characterized by a web-like pattern of fractures radiating outwards from a shallow shield.

asthenosphere

mechanically weak layer of the Earth underlying the rigid *lithosphere*, that flows under the constraints of convection currents and sometimes melts into *magma*. The asthenosphere comprises the crust and the uppermost section of the *mantle* and is roughly 100 km thick.

aureole deposit

debris-like deposit surrounding the volcano Olympus Mons on Mars. It is characterized by wave-like ridges and is attributed to mass wasting off the flank of the volcano.

breached (crater, cone)

asymmetric volcanic cone, open at one end in a horseshoe pattern. The shape is due to lava flowing out of the gap and preventing the build-up of a flank on that side.

breccia

impact breccia is a rock composed of fragments mixed and fused together by meteorite impacts. There are also volcanic and sedimentary types of breccia.

calc-alkaline suite

evolution of a *magma* that yields lava types increasingly richer in calcium and alkali metals (sodium and potassium). Typical of a *subduction* setting (island-arc and cordillera volcanoes).

calcite

mineral made of calcium carbonate. Limestone is a rock aggregate of calcite crystals.

caldera

circular or semi-circular depression at the center of a volcano, caused by the collapse of the surface over an emptying *magma* chamber.

carbonatite

rare lava composed of calcium or sodium carbonate crystals, that erupts at relatively low temperature and low viscosity.

clinopyroxene

silicate mineral of the *pyroxene* family, rich in iron and magnesium.

convection

circulation of hot material, caused by differences in density due to differences in temperature. Convection currents often take the form of vertical loops.

core

the central part of a celestial body. In the terrestrial planets, the core is principally made of iron and may or may not be molten.

corona

class of volcanoes on Venus, characterized by a central plateau, often surrounded by a moat and concentric faulting. Coronae are attributed to the spreading of a *hot spot* underground.

crust

the uppermost layer of a planet, which is derived from the *mantle* by igneous processes. The Earth has both continental and oceanic crusts that differ in thickness and composition.

cumulate

rock formed by the accumulation of dense crystals at the bottom or on the sides of a *magma* chamber. The rock *dunite* is a cumulate of *olivine* crystals.

dacite

lava type relatively rich in silica, containing *plagioclase* feldspar, quartz, and occasional hornblende crystals, typically found in island-arcs and other *subduction* settings (*calc-alkaline suite*).

diapir

   mass of rock that rises towards the surface because it is less dense than its surroundings.

differentiation

   evolution of a *magma* in which segregation of certain elements (absconded by minerals or flushed out of a system) can lead to a change in chemical composition of the *magma*, and thus to different types of lava. Different trends of differentiation are known as suites, such as the *alkaline* and *calc-alkaline suites*.

dike

   sheet of *magma* that rises vertically or sub-vertically through fractures in the crust (it is called a sill when it propagates horizontally). Dikes can stall before the surface or break out and feed volcanic eruptions.

dunite

   iron and magnesium-rich rock, composed principally of the mineral *olivine*.

effusion (effusive)

   magma that reaches the surface and flows as lava.

extremophiles

   life forms, such as algae and bacteria, that withstand extreme conditions of temperature, pressure, acidity, or salinity.

flood lavas

   lavas (often *tholeiite basalt*) that spread over large areas, in response to low viscosity and a high or sustained discharge rate. Also named *traps*.

fractional crystallization

   process by which different types of crystals form in a cooling *magma*, at different temperatures, and settle out of it, changing the chemistry of the remaining melt.

gabbro

   dark coarse-grained igneous rock. It is chemically equivalent to basalt, but crystallizes at greater depth.

graben

   down-drop of a section of crust between bounding faults, creating an elongated basin.

hawaiian eruption

   eruption characterized by lava fountaining, due to the relatively large amount of gas dissolved in the *magma*.

hematite

   iron oxide (Fe_2O_3) that occurs as gray crystals. Blood-red when powdered.

hornito

   conical mound that spatters small quantities of lava.

hot spot

abnormally hot part of the *mantle* that results in a buoyant *plume* of material that rises to the surface and feeds eruptions. On Io, hot spots instead designate abnormally hot areas at the surface, detected by infrared instruments.

hyaloclastite

breccia composed of small shards of volcanic glass, generated by the explosive reaction of hot lava with water.

ignimbrite

rock type composed of volcanic ashes compressed by their weight and fused together by their heat. Ignimbrites often constitute extensive sheets, deposited over large areas by a powerful eruption.

isotope ratio

ratio of different varieties of the same element, that have different masses (different numbers of neutrons), such as O_{16} and O_{18}.

komatiite

ultramafic lava (rich in iron and magnesium) that erupts at high temperature and low viscosity. Typical of a high-temperature, iron-rich *mantle*.

KREEP

chemical character of a class of lunar lavas, rich in potassium (K), rare Earth elements (REE) and phosphorus (P). These elements were concentrated in the last pockets of melt of the Moon's *magma* ocean.

lahar

mudflow rich in volcanic material, generated by heavy rainfall or the melting of surface ice by the eruption or the high heat flow of a volcano.

levee

raised embankment of a lava flow.

lithosphere

rigid upper layer of a planet, typically 70 to 150 km thick on Earth (except at mid-ocean ridges), composed of the crust and the uppermost part of the *mantle*. On Earth, the lithosphere is divided into tectonic plates.

maar

volcanic crater surrounded by a ring of debris, caused by the violent explosion of *magma* in contact with water. Often occupied by a lake.

mafic

which contains abundant magnesium and iron. Basalts are mafic.

maghemite

iron oxide mineral.

magma

liquid rock melt at high temperature, that turns to lava upon cooling and solidification. Magmas often contain dissolved volatiles that turn to gas, and carry in suspension crystals of forming minerals.

magma chamber

zone where rising magma comes in equilibrium with the surrounding rocks and stalls in place. Often represented as a sponge-like mush of solid rock and liquid melt, magma chambers can reach several kilometers in width and lead to chemical mixing and *differentiation* of *magma*.

magma ocean

global layer of *magma* that is thought to have stretched around the Moon, and perhaps the Earth, early in their histories. This global super-hot layer, hundreds of kilometers deep, would have created the early crusts of planets by *differentiation*.

mantle

middle part of a planet, between *core* and *crust*, at high temperatures and pressures. It is mineral and rich in silica, iron, magnesium, calcium and aluminum. Melting of the mantle at different depths yields different types of *magma*.

mare (maria)

lowlands of the Moon, covered with lava fields.

montmorillonite

volcanic clay of the smectite family, composed of hydrated silicate sheets, rich in iron, magnesium, sodium and aluminum.

neutral buoyancy zone

underground level at which *magma* stalls because its density matches that of the surrounding rock. *Magma chambers* often form in this zone.

nontronite

volcanic clay of the smectite family, composed of hydrated silicate sheets.

nuée ardente

type of *pyroclastic flow* that is gravity driven and contains hot gas and ash fragments.

olivine

green silicate mineral, rich in iron and magnesium.

pahoehoe

texture of a lava flow that boasts a smooth surface with "ropy" folds.

pali

escarpment or cliff on a Hawaiian volcano, due to the spreading and large scale collapse of the volcano's flank.

pancake dome

class of volcanoes on Venus with a circular plan form, steep sides and a flat top, probably caused by the extrusion of a large mass of frothy lava.

parasitic (cone, volcano)

> small cone or shield that grows on the flank of a large volcano.

patera (pl. *paterae*)

> class of volcanoes on Mars, Venus and Io, characterized by shallow flanks and a large central *caldera*. Paterae are shaped like overturned saucers.

pelean eruption

> type of eruption characterized by a flow of hot ash driven by gravity down the flank of a volcano, named after Mount Pelée in the French Caribbean.

peridotite

> typical *mantle* rock composed principally of *olivine*, *pyroxene* and feldspar.

phonolite

> type of viscous lava, rich in silica and containing feldspar, nepheline and *pyroxene* crystals. Belongs to the *alkaline suite*, typical of *hot spot* settings.

pillow lava

> type of lava that forms underwater with a glassy shell surrounding a crystalline heart. It owes its bulbous form to the surrounding water pressure and often stacks up in piles.

plagioclase (*feldspar*)

> silicate mineral rich in aluminum, that also contains calcium, sodium or potassium.

plinian eruption

> type of eruption characterized by a plume of ash rising into the atmosphere due to a high gas content. Named after the AD 79 eruption of Mount Vesuvius, described by Pliny the Younger.

plume

> current of hot and buoyant material that rises through the *mantle* of a planet, and can partially melt and erupt at the surface. Some plumes originate as deep as the *core–mantle* boundary. The expression of a plume at the surface is a *"hot spot"*.

pseudocrater

> crater and ring of debris in a lava field, due to a steam explosion blistering open a lava flow that creeps over water-logged terrain.

pyrite, pyrrhotite

> iron sulfide minerals.

pyroclastic flow

> flow of volcanic ash and gas, driven by gravity or propelled by a blast down the flank of an erupting volcano.

pyroxene

> volcanic silicate mineral, rich in iron, magnesium and calcium.

refractory

which melts at high temperature (opposite: *volatile*).

rhyolite

type of viscous lava very rich in silica (over 75%) that contains quartz, feldspar, amphibole and biotite. Rhyolite is the extreme *differentiation* product of basalt, or the melting product of continental crust.

rift zone

extension and thinning of the crust into down-dropped valleys that allow *magma* from the *mantle* to reach the surface. Locally on volcanoes, swarm of fractures that extend outward from a *magma chamber* and feed lateral eruptions.

rille (sinuous)

arcuate or meandering lava channel that often follows a pre-existing fault.

saponite

clay mineral.

scoria

pebble to fist-sized clump of lava with a rough texture and gas bubble holes, formed in lava fountains, eruptive clouds, or by the grinding of an advancing lava flow.

secondary crater

impact crater excavated downrange from a principal crater, and caused by a projectile thrown out from the main impact site (fragment of the projectile or of the displaced rock).

Shergottite

Sub-category of SNC "Martian" meteorites, named after the Shergotty meteorite. Basaltic composition.

shield

volcanic landform with shallow slopes, comparable in plan form to a warrior's shield, often crowned by a pit crater or a *caldera*.

SNC

rare class of meteorites named after the three main types Shergotty, Nakhla and Chassigny. They are now believed to be fragments of Mars, tossed into space by large impacts.

spatter cone

conical mound that spatters small quantities of lava.

spectral signature

Response of a gas, mineral or rock to light or other radiation. The resulting graph, as a function of wavelength, can be typical of a material and betray it, as would a signature.

spectrometer

remote-sensing instrument used to study the response of a target to light or other radiation.

stratovolcano

volcanic *shield* or cone composed of alternating layers of lava and ash.

strombolian eruption

type of eruption in which pockets of gas episodically burst at the top of a *magma* column, tossing clumps of lava above the vent.

subduction

in plate tectonics, process by which old and dense *lithosphere* (a crustal plate) bends downwards and sinks into the *mantle*. Volcanism often occurs behind the trench.

terra (pl. *terrae*)

highlands of the Moon.

tessera

ancient crumpled terrain of Venus.

tholeiite (*tholeiitic basalt*)

a form of basalt, relatively rich in silica, that is typical of mid-ocean ridges on Earth and has its source in the uppermost *mantle*.

tholus (pl. *tholi*)

hill in latin. Name attributed to a class of steeper-than-average Martian volcanoes.

trachyte

viscous lava, relatively rich in silica, that derives from alkaline *magma*, often in a *hot spot* setting.

trap

see *flood lavas*.

tube/tunnel (lava)

lava channel that builds up converging *levees* on its sides, to the point where the *crust* roofs over to form a tube or tunnel, insulating the flow from cooling and allowing it to flow over large distances.

tuya

volcanic landform, also named table mountain, that forms when *magma* comes to the surface beneath an ice cap and is molded by it into a steep-sided relief.

vacuole

cavity created in lava rock by the expansion and escape of gas bubbles.

viscosity

resistance to flow of a liquid or *magma*, which is principally a function of chemistry and temperature.

volatile
> which melts and vaporizes at low temperature (opposite: *refractory*).

vulcanian eruption
> type of eruption in which gas pressure mounts under an obstruction in the volcano's chimney and is released by an explosion that lofts heavy blocks out of the vent.

wrinkle ridge
> ridge at the surface of a lava field that probably forms from the buckling of the cooling lava.

Index

aa flow, 53, 55, 195, 250–51
Acala (lava field, Io), 289
accretion (planetary), 5–7, 64
Aci Castello (Etna), 50
Acidalia (plains, Mars), 192–93
Adirondack (rock, Mars), 154
Aidne Corona (volcano, Venus), *208, 224,*
Alba Patera (volcano, Mars), 127, *128,* 134,
    136, 139, 167, 169–70, 190–96, *190,*
    192–93, 195, Plate XI
Albor Tholus (volcano, Mars), *128,* 136–37,
    137, 197, *Plates XIV, XV*
Alcott (impact crater, Venus), 237
Aldrin, Buzz, 68–69
ALH 79001 (meteorite, Mars), 156
ALH 84001 (meteorite, Mars), 129, 156–58,
    161–63, *162*
alkali
    on Earth, 18, 55, 57, 294
    on Mars, 144
    on the Moon, 67
    on Venus, 210, 212
alkaline basalt, 41, 50–51, 60, 212–13, 245,
    Plate XIX
alkaline suite, 18, 41, 47–48
Alpha Regio (Venus), *208,* 215, *260–61,* 262
Alphonsus (crater, Moon), 66, 93
aluminous basalt, 71, 75, 84–85, 98,
    100–101, 108, 120
aluminum, 26, 69, 91, 108, 144, 148
Amazonian (era, Mars), 169–70, 191
Amazonis (plains, Mars), 172
Amirani (volcano, Io), *271,* 274, 281, 288,
    288, 300, 303–304, 319, *Plate XXIII*
Ammavaru (volcano, Venus), *208,* 265–268,
    265–66, 268
Amphitrites Patera (volcano, Mars), 138
andesite
    on Earth, 18, 29, 36, 60, 62–63, 141
    on Mars, 148, 150, 154, 157
    on Venus, 264
andesitic basalt, 18

anorthosite, 75, 101, 107
Apennine Bench Formation, Moon, 75, 84,
    93, 100–101
Apennine Mountains, Moon, 72–73, 73,
    99–100
Aphrodite Terra (Venus), *208,* 213, 215,
    217–18, *248–49,* 249, *Plate XVII*
Apollinaris Patera (volcano, Mars), *128, 146,*
    154, *Plate XVI*
Apollo landing sites (map), *65*
Apollo 11, 67–71
Apollo 12, 70–72, 244
Apollo 13, 72
Apollo 14, 72, 84
Apollo 15, 72–77, 81, *82–83,* 84–85, 93,
    97–99, 98–101, 102–103, 111
Apollo 16, 76, 95
Apollo 17, 76–83, 77, 79, 85, 87, 91,
    104–108, *104, 106–107, 109,* 111,
    Plates IX, X
arachnoid (Venus), 217, 219, 224–25, *226*
Aram Chaos (Mars), 151
archebacteria, 31
Arenal (volcano, Earth), 8
Ares Vallis (Mars), 148–49
Aristarchus (crater, Moon), 66, 93, *121,* 122,
    124
Aristarchus (plateau, Moon), 67, *86,* 93,
    120–26, *121, Plate VIII*
armalcolite, 84
Armstrong, Neil, 68–69
Arsia Mons (volcano, Mars), *128,* 130, 132,
    134, 167, 184–89, *184, 187–89,*
    279
Artemis Corona (Venus), *208,* 217
Ascraeus Mons (volcano, Mars), *128,* 130,
    132, 134, *135,* 167, 184
ash (fall, layer)
    on Earth, 19, 22–28, 52–53, 61–62,
    Plate VI
    on Io, 291, *291,* 311, 330 *and see* Io:
    plume, fallout ring

on Mars, 134, 136–37, 140–42, 170,
        185–86, 194–201, 203–205, *204*
    on the Moon, 66, 76, 119–20, *Plate VIII*
    on Venus, *233*, 233–34, 250, 264
Aso, Mount (volcano, Earth), 66
asthenosphere, *9*, 10, 293–94
Aten (volcano, Io), 278, 281, 285, 289
Atla Regio (Venus), *208*, 221, *223*, 230, 236,
        249–50
atmosphere, *see individual planets*
aureole deposit (Mars), 43, *175*, 175–76

Babbar Patera (volcano, Io), *271*, *305*, *309*
bacteria
    on Earth, 30–31, 163–64
    on Mars, 129, 156–57, 161–64
Balch (impact crater, Venus), *246*
balloon (exploration by), 191–93, 244–47, 250
Barnacle Bill (rock, Mars), 148, *149*
basalt
    on Earth, 18, 29, 39, 41
    on Io, 290, 294, 300, 321, 325, 327
    on Mars, 141, 148, 150–51, *153*, 154,
        156–58
    on the Moon, 67, 69–71, 74–76, 78,
        109–110, 125
    on Venus, *211*, 212–13, 230, 264, 267
basaltic andesite
    on Earth, 29
    on Mars, 148–49
base surge, 26, 140
BAT zone (Venus), 236
Bean, Alan, 70–71, *71*
Bell Regio (Venus), 236
Beta Regio (Venus), *208*, 210, 215, 218, *218*,
        221, 223, 236, 242–46, *242–43*, *245*
Biblis Patera (volcano, Mars), *128*, 134, *135*
blast (volcanic), 22, 60
bomb, *see* lava bomb
Bosphorus Regio (Io), *270*, *286*, 319, *326*, 328
breached cone, 113–14, *114*, 262
breccia, *see* impact breccia
bubble, *see* gas bubble

calc-alkaline suite, 18, 60
calcite, 145
calcium, 54, 83, 144, 152
caldera
    on Earth, 16–17, *17*, 19, 35–38, *40*,
        41–44, *43*, 48, 53, 55, 60, *62*, 62–63,
        279, *Plates IV, V*

on Io, 274, *278*, 279–81, *280*, 285, 287,
        290–92, 295, *296*, 298–308,
        301–302, *310*, 311, *312*, 312–13,
        314, 316, *317*, 318–19, 323–24, *326*,
        327–28
    on Mars, 130, *131*, 132, *133*, 138, *139*,
        140, *141*, *146–47*, 175–76, 181–83,
        180–82, *184*, *185*, 187, 187–89, 191,
        194–95, *195*, 197, 199–201, *200*,
        202, 203, 205–206, *206*, 279,
        Plates XII, XIII, XIV, XVI
    on Venus, 215–17, *216*, 222, *222*, *240*,
        244, 247, 250–52, 256–57, 267
Callisto, 272, 287, 302
Camaxtli Patera (Io), *301*
carbon dioxide
    in atmosphere, 156, 165, 168, 170,
        209–10, *214*, 239, 245
    in melt, 18, 20, 22, 55, 60, 168, 233
    in plume, 24, 165
carbon monoxide, *82*, 114
carbonate
    on Mars, 145, 161–62, *163*
    on Venus, 240
carbonatite, 54–58, 54, *54*, *58*, 230
Ceraunius Tholus (volcano, Mars), *128*, 134,
        135, 159–60, *160*, 192
Cerberus (plains, Mars), 160
Cernan, Gene, 78–80
Cerro Galan (caldera, Earth), *Plate V*
Chaac Patera (volcano, Io), 279, 300
Chain of Craters (Hawaii), 45–46
Challenger (Space Shuttle), 286
channel (lava), *see* lava channel
channel (water)
    on Mars, 136–38, *139*, 148, 152, 154,
        169–70, 197, 200, *202*, 203–205, *204*
Chassigny (meteorite, Mars), 154–55, *155*,
        157–58, 160
Chicxulub impact crater, 30
chlorine, 20, 27, 144
Chryse (plains, Mars), 143, 148
cinder cone, *see* cone
clay (on Mars), 145, 162, 205
Clementine (spacecraft), 85, *88*, 118
climate (eruptions and), 24–26, 28, 170
clinopyroxene, 150, 157
cloud ring, *52*
Cobra Head (Moon), *121*, *124*, 124
collapse (of volcano flank), 43, 48, *49*, 52–53,
        60, 63, *147*, 175–77, 227, *Plate IV*

Columbia River Plateau, 11, 28–29, 164, 265
cone (volcanic)
  on Earth 48, 51–53, 52, 57
  on Mars, 171–72, 172, 196
  on the Moon, 94, 96, 105, 111, 113, 114,
    115
  on Venus, 219, 221, 250
Conrad, Peter, 70–71
continental drift (Earth), 12
convection (in mantle), 12, 15, 138, 235,
    239, 293
cooling (of climate), 24–26, 28
Copernicus (crater, Moon), 70
core
  of Earth, 7–10, 9
  of Mars, 165
  of Moon, 88–90
  of Venus, 239
core-mantle boundary, 10, 29
corona, 216–17, 219, 221, 224–27, 224,
    255–60, 255–56, 258–59
crater, see caldera, impact crater, vent
crust
  of Earth, 9, 10, 12, 14–15, 167
  of Io, 281, 293–94, 308, 316
  of Mars, 167, 194
  of the Moon, 87–88, 90–91
  of Venus, 219, 232, 238–241
cryptomaria, 87
Culann (volcano, Io), 271, 288, 292, 298, 319
cumulate, 155, 156–57, 160, 162

D-prime zone, 10
dacite, 18, 36–37, 60, 141, 227
Dali Chasma (rift, Venus), 249
Danu Montes (Venus), 218
Danube Planum (Io), 306, 308, 310
Dao Vallis (Mars), 202, 204, 204–205
Dar Al Gani (meteorite, Mars), 158
dark halo crater (Moon), 88, 105
dark mantle deposit, Moon, 94
Dazhbog Patera (Io), 303
debris flow, 43, 175, 177, 264
Deccan Traps, 28–30, 93, 136
Descartes crater (Moon), 76, 95
Devana Chasma (rift, Venus), 243–47, 243,
    246
diamond, 55
diatreme, 55
differentiation (of magma), 18, 36–37, 41, 52,
    144, 148, 254, 294

dike
  on Earth 38, 53, 55
  on Mars, 129, 166, 171, 194
  on Venus, 224–25, 232, 250–51
discharge rate (of ash, lava), 23, 42, 72, 74,
    93–94, 96, 101, 140, 238, 269, 292,
    300–301, 307
dome (volcanic)
  on Earth, 19, 62, 62, 141, 227, Plates IV,
    V
  on Io, 297
  on Mars, 187, 189
  on Moon, 111, 113
  on Venus, 217, 224, 224, 227–29, 228,
    235, 250, 252, 254, 260–64, 260–61,
    263–64
Duke, Charlie, 76, 95
dunite, 91, 108, 155, 157, 160

Earth
  accretion, 5–7
  asthenosphere, 9, 10
  ash falls, 19, 22–28, 52–53, 61–62,
    Plate VI
  calderas, 16–17, 17, 19, 35–38, 40,
    41–44, 43, 48, 53, 55, 60, 62, 62–63,
    279, Plates IV, V
  cones, 48, 51–53, 52, 57
  core, 7–10, 9
  creation and heat flow, 4–10
  crust, 9, 10, 12, 14–15
  domes, 19, 62, 62, 141, 227, Plates
    IV, V
  eruptions, 8, 18, 21–29, 35, 41–42,
    49–53, 55–58, Plates I, II, III, VI
  hot spots, 11, 15, 18, 29, 41, 47–48, 50,
    138
  internal structure, 9, 9–10
  lava field, flow, 15, 18, 21, 26, 38–39,
    41, 44–46, 48–53, Plates I, II, III
  magma ocean, 6
  magnetic field, 10
  mantle, 9, 10–12
  mid-ocean ridges, 5, 13, 13, 15, 30–31,
    31, 34–35, 165–67, 217–18
  plate tectonics, 12–15
  plume (eruptive), , 21, 22, 24, 25, 53,
    Plates IV, VI
  seamounts, 13
  shield volcanoes, 48, 51
  topographic map, 5

East African Rift, *8*, *14*, *17*, 55–56, 171, 218, 244–45
East Pacific Rise, *31*
eclipse (on Io), 290
Eifel (volcanoes, Earth), 213
Eistla Regio (Venus), 256
Elysium Mons (volcano, Mars), *128*, *136–37*, 137, 197, *198*, *Plate XV*
Elysium (plains, Mars), *159*, 172–73
Elysium rise (Mars), 136–37, *136–37*, 158–60, 197, *Plates XIV, XV*
Emakong (volcano, Io), 298, 303, 326–30, *326–27*, *329*
End-Cretaceous crisis, 27–28, 30
End-Permian crisis, 28, 30
endogenic dome, 227, 261
Eolian Islands, 22, 47
Erebus (volcano, Earth), 8
erosion
    on Earth, 225
    on Mars, 189, 193
    on Venus, 209
Erta Ale (volcano, Earth), *5*, 8, *17*
eruption
    on Earth, 8, 18, 21–29, 35, 41–42, 49–53, 55–58, *Plates I, II, III, VI*
    on Io, 273–74, *273*, 277, *278*, 281, 288–92, *291*, *293*, 295–308, *306*, 311, 313, 316, *Plates XXI, XXII*
    on Mars, 129, 136–37, 140–42, 170–71, 176, 181–83, 186, 189, 191, 197–201
    on the Moon, 74, 76, 93–96, 105, 114, 119
    on Venus, 209, 213–14, 232–34, 238–39
    hawaiian, 22, 45, 53, 140, 232, 234, *299*, *Plate I*
    pelean, 23, 60, 63, 232
    plinian, *21*, 22, 55, 58, 60, 62, 140–42, 197, 199–200, 232–34
    strombolian, 22, 53, 55, 57, 232, 234
    vulcanian, 22, 23, 55, 58, 234
eruption rate, *see* discharge rate
Etna, Mount (volcano, Earth), *5*, *6*, 8, 26, 46–53, *47*, *49–50*, *52*, 124, 238, 247
Europa, 272, 276, 287, 304
europium, 90
Evans, Ron, 78, 85
excentric eruption, 51
extremophile (bacteria), 30, 63

fallout ring (Io), 274, 278, 285, 289, 303, *305–306*, 307, *309*, 313, 318
famine, 26–27
far side, of the Moon, 67, 113
feldspar, 51, *83*, 83, 90, 101, 107, 125, 150, 153–54
fissure (eruptive)
    on Earth, 13, 34–35, 41
    on Io, 274, *278*, 291–92, 297–98, *312*, 313, 318
    on Mars, 137, 140, 171, 181, 188–89
    on Moon, 114
flood basalt, 29, 66, 72, 93, 265, 281
flood channel, *see* channel
flow, *see* lava flow
fluorine, 20, 27
fossil, microfossil
    on Mars, 161, *163*, 172, 176, 203
fractional crystallization, 18, 157
fracture belt (Venus), 221, 229, 259, *Plate XX*
Fra Mauro (hills, Moon), 72, 84
Franklin, Benjamin, 24
fumarole, 108, 283

Galapagos Islands, 148, 263
Galileo Galilei, 64–65
Galileo, 85, 286–290, 292–304, 306, 318, 323, 326, 328
Ganymede, 272, 287, 302
gas
    in atmosphere, 24–27, 239
    in magma, 18–27, 119–20
gas bubble, 20–22, 115, 140, 142, 156, 232–33, 252
Genesis rock, Moon, 75
geothermal energy, 36, 38
    on Mars, 129
glacial deposits (on Mars), 176, 186, *188*
glass, volcanic
    on Moon, 69, 76, 79–82, *81*, 94, 101, 105, 108–109, *109*, 119–20, 122, 125
    on Mars, 205
graben, 105, *136–37*, *181*, 187, 191, 194
gravity, effect on eruption, 94, 114, 140, 142, 199–200, 274
Greeley, Ron, 112
greenhouse effect
    on Earth, 24
    on Mars, 170
    on Venus, 207, 209, 239–41

Grimaldi crater (Moon), 93
groundwater (on Mars), 136, 163, 167–68, 173, 205
Gruithuisen domes (Moon), 95, 116–120, *116–17, 119*
Guinevere (plains, Venus), *233*
Gula Mons (volcano, Venus), *208*, 223, 255–60, *255–56, Plate XX*
gully (on Mars), 205
Gusev (crater, Mars), 152–54

Hadley, Mount (Moon), 100
Hadley Rille, Moon, 72–76, *73, 75*, 94, 97–99, 97–103, *102–103*, 111, 328
Hadriaca Patera (volcano, Mars), *128*, 138, 202–206, *202, 204, 206*
Halemaumau (pit, Kilauea), 17, *40*, 42, *43*, 44
Harbinger domes (Moon), 67
Hawaii volcanoes, 11, *11*, 22, 40–46, 74, 80, 148, 176, 300
hawaiian (eruption), 22, 45, 53, 140, 232, 234, *299, Plate I*
Hawaiian Islands (chain), 40–41, 139
hawaiite, 41
Head, Jim, 251, *252–53*
heat flow
    on Earth, 4–10, 219
    on Io, 276, 278–79, 304, 306–308, 312, 316, 318, 327
    on Mars, 173
Hecates Tholus (volcano, Mars), *128*, 137, 141, 196–201, *Plate XV*
Hellas (basin, Mars), 136, 138, 141, *172*, 203–204
hematite, 151–52, 167
Hermann Formation (Moon), 113
Herodotus crater (Moon), *86, 121*, 121, 124
Hesperian (era, Mars), 136, 169–70, 191, 197
highland volcanism (Moon), 76
Hildr channel (Venus), 230
homo sapiens, 26
hornito, 55, 57, *and see* spatter cone
horseshoe crater, *see* breached cone
hot spot
    on Earth, 11, 15, 18, 29, 41, 47–48, 50, 138
    on Io, 276–79, 290, 292, 303–304, 307, 312–13, 318–19
    on Mars, 139
    on Venus, 209, 213, 219, 225, 235–36, 244, 249, 256–57

hot spring, 36, 63, 176, 183
Hubble Space Telescope, 307, 323
Humphrey (rock, Mars), *153*, 154
Hverfjall (volcano, Earth), 37
hyaloclastic ridge, 39
hyaloclastite, 50
hydrothermal circulation,
    on Earth 30–32, 36, 63, 163–64
    on Mars, 129, 151, 161–63, 173, 203, 205
    on the Moon, 80
hydrothermal vent, 30–31, *31*, 129, 163

Iceland, 24, *34*, 96, 148, 163, 171
iddingsite, 162–63
Idem-Kuva Corona (volcano, Venus), *208*, 227, 255–60, *255, 258–59*
ignimbrite 8, 19, 60
Imbrium (basin, impact), 74, 76, 93, 101, 117–18, 122
Imbrium, mare, *see* Sea of Rains
impact basin, 70–71, 83
impact breccia, 69, 72, 76, 78, 91–92, 107
impact crater
    on Mars, 130, 132–33, 152, 159–60, 165, *202*
    on the Moon, 66, 79, 82, 86
    on Venus, 237–38, *237*, 251, 257
impact melt, 91
incompatible element, 90
inflated lava flow, 251, 318
Io
    ash falls, 291, *291*, 311, 314, 330, *and see* fallout rings, plumes
    brightenings, 276, 287, 298, 313
    calderas, 274, *278*, 279–81, *280*, 285, 287, 290–92, 295, *296*, 298–308, *301–302, 310*, 311, *312*, 312–13, *314*, 316, *317*, 318–19, 323–24, *326*, 327–28
    crust, 281, 293–94, 308, 316
    colors, 272, 281–82
    domes, 297
    eruptions, 273–74, *273*, 277, *278*, 281, 288–92, *291, 293*, 295–308, *306*, 311, 313, 316, *Plates XXI, XXII*
    fallout rings, 274, 278, 285, 289, 303, *305–306*, 307, *309*, 313, 318
    heat flow, 276, 278–79, 304, 306–308, 312, 316, 318, 327
    hot spots, 276–79, 290, 292, 303–304, 307, 312–13, 318–19

infrared, 276, 286–87, *286*, 290, 303–304, 307, 312–13, 318, 320

lava flows, *280*, 284–85, *288*, 289–92, 295, 297–98, 300–301, *301*, 303, 308, 310, 313, *317*, 318–21, *319–20*, *322*, 323–30, *326–27*, *329*, *Plates XXI, XXIII, XXIV*

lava lakes, 281, 287, 292, 295, *296*, 300, 303, 306–308, 311, *312*, 313, *314*, 316, 327

magma, 290–91

magnetic field (Jupiter's), 287, 295, 297, 304, 309–310

mantle, 290, 293–94

map of, *271*

orbital parameters, 275–76

outbursts, 313, 316

plumes, 273–74, *273*, 276–77, *278*, *282*, 288–90, 292, 294, 297, 300, 302–303, 305, *306*, 307–308, 311, 313, *315*, *317*, 318–19, 321, 323–24, *325, Plates XXI, XXII*

ring of fire, 286, 319, 326

shields, 279, *280*, 297, 323

tectonics, 300, 306

tidal stress, 275–76, 292–93

iron, 10, 17–18, 45, 66, 69, 83–84, 108–109, 118, 122, 125, 143–45, 148, 150–51, 162, 290, 298, *Plate VIII*

Irwin, James, 73–75, *75*, 84, 98, 100

Ishtar Terra (Venus), *208*, 215, 217–18, *Plate XVII*

Isidis (impact basin, Mars), 150

island arc, 47, 59–60, 217

Izu-Oshima (eruption), 22

jökullaup, 24, 199

Jovis Tholus (volcano, Mars), 134, *135*

Juan de Fuca (ridge, Earth), 13

Jupiter, 270–72, 275–76, *275*, 286–87, 290, 292, 297, 303–304, 309, 314, 317–18, 320, *325*

K-T boundary, 28, 30

Kamchatka volcanoes, *5*, 25, *Plate VI*

Karoo Traps, 30

Katmai (volcano, Earth), 8, 19

Keddie, Susan, 251, *252–53*

Kanahekili (lava field, Io), 289

Keck Observatory, 304

Keanakakoi ash (Hawaii), *42*, 44

kieserite, 145

Kilauea (volcano, Earth), *5*, 8, 17, 22, 40–46, *40, 42–43, 45*, 93, 201, 206, 238, *296*, 300–301, 328, *Plates I, II, III*

Kilauea Iki (volcano, Earth), *43*, 44–45

komatiite, 230, 267, 290, 294, 298, 300, 321, 325, 327

Krafla (volcano, Earth), *5*, 33–39, *33–34*, *37–38*

Krakatao (volcano, Earth), *5*, 8, 24–25

Kreep basalts, 85, 101, 118, 120

laccolith, 44

Lada Terra (Venus), 266

Lafayette (meteorite, Mars), 158, 162–63

La Garita (caldera, Earth), 19

lahar, 24, 199

Laki (volcano, Earth), 24, 26–27, 292, 301

Large Igneous Province, 29

Lassen Peak (volcano, Earth), 227

Lastarria (volcano, Earth), 283

lava bomb, *23*, 62, 142, 234

lava channel
    on Earth, 44, 51, 55, 282
    on Io, 298, 328–30, *329*
    on the Moon 67, 72–76, *73*, *75*, *86*, 96, 100–101, 111, *112*, 113–15, *123–25*, 124
    on Mars, 177, 179, 186, 188
    on Venus, 217, 219, 229–31, *231*, 241, 265–69, *266, 268*

lava field, flow
    on Earth, 15, 18, 21, 26, 38–39, 41, 44–46, 48–53, *Plates I, II, III*
    on Io, *280*, 284–85, *288*, 289–92, 295, 297–98, 300–301, *301*, 303, 308, 310, 313, *317*, 318–21, *319–20*, *322*, 323–30, *326–27*, *329, Plates XXI, XXIII, XXIV*
    on Mars, 129, 132, 134, 140, 143, *143*, *151*, *159*, 160, 168–69, 175–83, *177–78*, 184–88, *187*, *189*, 191, 194–95, 198, 203, *206, Plate XVI*
    on Moon, 66–76, 86, *92*, 93–94, 96, 113, 122
    on Venus, 210–13, *211*, 215–17, 219–21, *216*, 227, 229–30, 232, *235*, *237*, 244–47, 249–51, *252–54*, 254–57, *255–256*, 259, 265–269, *265–66, 270*

lava fountain
    on Earth, 21–22, 35, 45, 51, 53, 81, 94, *299, Plate I*

lava fountain (*cont.*)
  on Io, 292, 298, *299*, 301–302, 304, 311
  on Mars, 140, 142, 201
  on Moon, 76, 81, 94, 101, 108, 125
  on Venus, 234
lava lake
  on Earth, *17*, 42, 44–45, *45*, 58
  on Io, 281, 287, 292, 295, *296*, 300, 303,
    306–308, 311, *312*, 313, *314*, 316
  on Mars 182–83
  on Venus, 247
lava needle, *61*
lava plateau, *29*
lava suite, 18
lava tube (tunnel), 39, 42, 46, 55, 74, 96, 111,
    113–15, *115*, 231, 282, 295, 297,
    318, 330
Lavinia Regio (Venus), *237*, 266
leucobasalt, 18
levee (of lava channel), 51, 53, 58, 74, *177*,
    179, 186, 268, 282
life
  on Mars, 129, 144, 151–52, 161–64
  origin of, 30–32
Lincoln-Lee scarp (Moon), *106*, 110
lithosphere, 225
Loki (volcano, Io), *271*, 274, 277–78, *278*,
    281, *286*, 287–88, 295, 303, 312–17,
    312, *314–15*, 324
Long Valley (caldera, Earth), 19
Los Angeles (meteorites, Mars), 157
Ludent (volcano, Earth), 37
Luna landing sites (map), *65*
Luna, 66, 82–83
Lunar Orbiter, 67, *111–12*, *114*, *117*
Lunar Prospector, 85–87, 118
lunar rover, 73, 75

maar, 23, *35*, 38, 57, 186
Maasaw Patera (volcano, Io), *280*
Maat Mons (volcano, Venus), *208*, 222, *223*,
    235, 238, 249–50, *249*
mafic (lava), 143, 145
Magellan, 219, 238
maghemite, 145
magma
  in general, 10, 15–23
  ascent of, 15–16, 35–36, 41, 138, 144,
    194, 221, 224–25, 232, 306
  on Earth, 10, 15–23, 35–36, 38, 224–25,
    294

  on Io, 290–91, 306
  on Mars, 144, 168, 194
  on Venus, 221, 232
magma chamber (reservoir)
  on Earth, *16*, 16–19, 36, 38, 41–42, 48,
    51–52, 148
  on Mars, 140, 156–57, 160, 177, 181,
    183, 187, 201
  on the Moon, 108
  on Venus, 215, 221, 229, 232, 250, 252,
    260
magma ocean, 6, 85, 89–91, 101, 107, 294
magnesium, 17–18, 76, 83–84, 108, 143–44,
    150, 152, 157, 290–91, 298, 307,
    321
magnetic field
  on Earth, 10, 165–66
  on Mars, 165–67, *166*
  on Io (Jupiter's), 287, 295, 297, 304,
    309–10
magnetite, 161
manganese, 31
mantle
  of Earth, 9, 10–12
  of Io, 290, 293–94
  of the Moon, 88
  of Venus, 239–40
Marduk (volcano, Io), 274, 288
mare, maria (lunar sea), 64, 66–67, 92–94
Mariner 4, 130
Mariner 5, 209
Mariner 6, 130
Mariner 7, 130
Mariner 9, 130, 132
Marius Hills (Moon), 67, 96, 110–16,
    111–12, *114*
Mars
  age of volcanism, 127, 129, 132–34,
    136–37, 158–61, 176, *178*, 183, 185,
    188, 191, 197, 200, 203
  ash clouds and fields, 136–37, 140–42,
    170, 185–86, 194–201, 203–205, *204*
  atmosphere, 129, 140, 168–70
  calderas, 130, *131*, 132, *133*, 134, 138,
    139, 140, *141*, 146–47, 175–76,
    181–83, *180–82*, *184*, *185*, 187,
    187–89, 191, 194–95, *195*, 197,
    199–201, *200*, *202*, 203, 205–206,
    206, Plates XII, XIII, XIV, XVI
  channels, 136–38, *139*, 148, 152, 154,
    169–70, 197, 200, *202*, 203–205, *204*

cones, 171–72, *172*, 196
core, 165
crust, 167, 194
domes, *187*, 189
eruptions, 129, 136–37, 140–42, 170–71, 176, 181–83, 186, 189, 191, 197–201
fossils, 129, 161, *163*, 170, 203
hot spots, 139
ice caps, 169–70
impact craters, 130, 132–33, 152, 159–60, 165, *202*
lakes and seas, 151–52, 168–70, 176–77, 194
lava flows, 129, 132, 134, 140, 143, *143*, *151*, *159*, 160, 168–69, 175–83, *177–78*, 184–88, *187*, *189*, 191, 194–95, 198, 203, *206*, *Plate XVI*
life on, 129, 144, 151–52, 161–64
map of, *128*
meteorites from, 127, 134, 154–58, *155*
northern plains, 150, 166, 169, 176, 193–94
orbital parameters, 129
plumes (eruptive), 142, 201
shields, *131*, 132, 134, 137, 150, 169, *174–75*, 175–81, 184–86, *184*, *190*, 190–91, *192*, 197, 199, *Plates XII, XVI*
size, 130
soil, 143–45, 148–49, 151
Mars Exploration Rover (MER), 127, 152–54
Mars Express, 127, 165
Mars Global Surveyor, 127, 132, 149–51, 164–66, 168, 176, 198
Mars Odyssey, 127, 165
Mars Pathfinder, 148–50, 154
mascon, 87
mass extinctions, 27–30
Masubi (volcano, Io), 274, 288, *Plate XXII*
Mattingly, Ken, 85
Maui (volcano, Io), *271*, 274, 288, *288*, *Plate XXIII*
Mauna Loa (volcano, Earth), 40–41, 132, 283
Mauna Ulu (volcano, Earth), 45
Maxwell Montes (Venus), *214*
McKenzie (dike swarm, Earth), 225
melt, melting, 10, 15, 18, 20
Merapi (volcano, Earth), *5, 20*
Meteor Crater, 66
meteorites, volcanic origin of, 65

from Mars, *see* Mars meteorites, *and* SNC
methane, 164
microfossil, *see* fossil
Mid-Atlantic ridge, 13, 34–35
mid-ocean ridge, *5, 13*, 13, 15, 30–31, *31*, 34–35, 165–67, 217–18
minerals
    on Earth, 84
    on Mars, 144–45, 148
Mitchell, Edgar, 72, 84
Monti Rossi (Etna), 51
Monti Silvestri (Etna), 52
montmorillonite , 145
Moon
    ash falls, 66, 76, 119–20, *Plate VIII*
    channels, 67, 72–76, *73, 75*, 86, 96, 100–101, 111, *112*, 113–15, *123–25*, 124
    cones, 94, 96, 105, 111, 113, *114*, 115
    core, 88–90
    craters, 65–66
    crust, 87–88, 90–91
    creation, 6, 89–91, *89*
    domes, 111, 113
    eruptions, 74, 76, 93–96, 105, 114, 119
    exploration, 66–87
    impact craters, 66, 79, 82, 86
    internal structure, 88–89
    lava flows, 66–76, 86, *92*, 93–94, 96, 113, 122
    magma ocean, 85, 89–91, 101, 107
    mantle, 88
    maria (seas), 64, 66–67, 92–94
    remote-sensing, 85–88
    shields, 95–96, *95*, *111–12*, 113, 124
moon quake, 88
Morabito, Linda, 272
Mount, *see* Etna, Saint-Helens, etc.
mudflow, 26, 61, 63, 198–99
mud pot, 36, 38–39
mugearite, 41
Mylitta fluctus (lava field, Venus), *208*, 229–30, 238, 265–270, *266, 270*
Myvatn (Lake, Iceland), 34, 37

Nakhla, nakhlite (meteorite, Mars), 155, 157–58, 162–63
Natron (Lake, Earth), 56–57
Namaskard (mud pots), 38
near side (of Moon), 67, 87, 113

nephelinite, 55, 57
neutral buoyancy zone, *16*, 16
New Zealand, 163
Ngorongoro (volcano, Earth), 56
Noachian (era, Mars), 165, 168, 170
nontronite, 145
norite, 107
nova, *see* stellate fracture center
nuée ardente, 23, 26, 60–61
Nyiragongo (volcano, Earth), 26
Nyos, lake, 26

Ocean of Storms, 70–72, *71*, 84, *86*, 93, *95*,
 95–96, *111*, 112–13, *116*, 117–18,
 122
Ol Doinyo Lengai (volcano, Earth), *5*, 54–58,
 54, 56, 58
olivine, 17, 45, *83*, 84, 88, *91*, 108, 150, *153*,
 154–55, 157
olivine basalt, 41, 71, 76, *82–83*, 102, *153*,
 154
Olympus Mons (volcano, Mars), 43, 53, 127,
 128, 129, *131*, 131–32, 134, *135*,
 139, *141*, 160, 167, 169, *174–75*,
 175–83, *177–78, 180–82*, 184, 190,
 Plates XI, XII, XIII
Opportunity, 151–52
orange soil, Moon, 79, 80–82, *81*, 94, 105,
 108–109, *109, Plates IX, X*
ore (mineral), on Mars, 129, 173
Orientale, *mare* (Moon), 91
orthopyroxene, 157, *162*, 291, 307
Ovda Regio (Venus), *235*
Ozza Mons (volcano, Venus), 221–22, 249,
 249

Pacific basin, volcanism, 14
PAH (Polycyclic Aromatic Hydrocarbon),
 161
pahoehoe lava, *33*, 39, 55, 58, 195, 250, 281
pali (scarp, Hawaii), 43, 46
pancake domes (Venus), *208*, 227, *228*, 229,
 260–64, *260–61, 263–64*
parasitic cone, 48, 57, 221, 251, *253*, 262
Parana-Edenteka Traps, 28, 30
partial melting, 18
patera (pl.: paterae) (volcanic construct)
 on Io, 279, *280*, 281
 on Mars, 134, 141
Pathfinder, see Mars Pathfinder
Pauahi (lava lake, Hawaii), *296*

Pavonis Mons (volcano, Mars), *128*, 130,
 132, *133*, 134, *135*, 167, 184
Peale, Stanton, 275
Pele (volcano, Io), *270*, 274, 277, 285, 288,
 293, 295, *296*, 300, 303, 305–311,
 305–306, 309–310, 324
Pele's hair, 81
pelean (eruption), 23, 60, 63, 140–42, 232
Pelée, Mount (volcano, Earth), *5*, 26–27,
 59–63, *59, 61–62*, 227
peridotite, 18, 157
Phlegrean craters (Earth), 65
Phoebe Regio (Venus), *211*
phonolite, 18, 55, 57
phosphorous, 85, 91
phreatic (eruption), 63
Piano del Lago (Etna), 48, *50*, 53
picritic basalt, 45
Pillan (volcano, Io), *271*, 288–92, *291*, *293*,
 295, 297, 300, 308, *Plate XXI*
pillow lava, 34, 50, 171
Pinatubo, Mount (volcano, Earth), 25, 199,
 213
Pioneer Venus, 213–15
pit (crater), *193*, 194, 221, 233, 251, 261, 263,
 297, 300, *301, Plate XIII*
plagioclase, *83*, 83, 90, 150, 153–54, 157
plate tectonics
 on Earth, 12–15, 20, 138
 on Mars, 166–67
 on Venus, 209, 217–19, 235, 239
plinian (eruption), *21*, 22, 55, 58, 60, 62,
 140–42, 197, 199–200, 232–34
plume (eruptive),
 on Earth, *21*, 22, 24, *25*, 53, *Plates IV*,
 on Io, 273–74, *273*, 276–77, *278, 282*,
 288–90, 292, 294, 297, 300,
 302–303, 305, *306*, 307–308, 311,
 313, *315, 317*, 318–19, 321, 323–24,
 325, Plates XXI, XXII
 on Mars, 142, 201
 on Venus, 214, 234
plume (mantle)
 on Earth,, 12, 15, 29, 34, 41, 138, 219
 on Mars, 189
 on Venus, 219, 221, 225, 235, 249,
 259–60, 262
potassium, 18, 67, 69–70, 85, 91, 118, 144,
 152, 210, 212, 245, 284
pressure
 effect on eruption, 20, 22, 60, 94,

140–42, 209, 214, 229–30, 232–34, 253, 262, 274
effect on melting, 9–10
Prometheus (volcano, Io), *271*, 274, 281, *286*, 288–89, 292, 297, 300, 317–321, *317*, *319–20*, *Plate XXI*
pseudocrater, 37, 171–72, 297
pumice, 60, 63
push-up ridge, 39
Pu'u O'o (volcano, Earth), *11*, 16, 41–42, 46, *Plate I*
Pu'u Pu'ai (volcano, Earth), 45
Puys, chaîne des (volcanoes, Earth), 18
pyrite, 164, 258, 298
pyroclastic (deposit, flow), *see also* ash flow
  on Earth, 23, 49, 52, 60–61, 63
  on Io, 291
  on Mars, 136, 140–42, 179, 198, 203
  on Moon, 94, 120
  on Venus, 261, 264
pyroxene, 71, *83*, 84, 88, 150, 154, 157
pyroxene basalt, 76, 102
pyroxenite, 160
pyrrhotite, 167, 258
quartz basalt, 71

Ra Patera (volcano, Io), *271*, 281, 284, 322–25, *322, 324*
radial dike swarm, 225
radioactive decay, 7, 70
radon, 122
rare earth elements, 85
refractory elements, 70, 83
regolith (lunar soil), 119, *Plates VII–X*
resurgent dome, 19, *Plate V*
Rhea Mons (volcano, Venus), 215, 243–46
rhyolite, 18, 36, 38–39, 141, 227
ridge belt (Venus), *266*, 267–69, *270*
rift zone
  on Earth, 16–17, 19, 34–35, 41, 45–46, 48–49, *299*
  on Io, 291, 305, 307–308, *310*, 311, 314
  on Mars, 134, 167, 184–86
  on Venus, 215–18, *218*, 221–23, 242–49, *242–43, 248–49*, 256–59, *Plate XVII*
rille, *see* sinuous rille
ring fault, 19, 100, 117–18, 187, 201, 216–17, 221, 250–51, 257, 267
ring of fire (Io), *286*, 319, 326
rootless vent, *see* pseudocrater

Roza flow (Earth), 28–29
Rumker Hills (Moon), *95, 96*

Sacajawea Patera (volcano, Venus), *208*, 216, *216*
Saint Helens, Mount (volcano, Earth), *5*, 8, *21*, 227, 291, *Plate IV*
Saint-Pierre (town, Earth), *59*, 60–63, *61*
Sakurajima (volcano, Earth), 8, 22
Sapas Mons (volcano, Venus), *208*, 223, *223*, 248–254, *248–49, 252–54*, *Plate XVIII*
saponite, 145
scarp, 43, 46, 175–76, *195*, 206, 284
Schiller-Shickard basin (moon), 87
Schmitt, Harrison, 77, *77*–80, 82, *Plate IX*
Schrödinger basin (Moon), *88*
Schroeter's valley (Moon), *86*, 120–26, *121*, *123–25*
scoria, 81, 94, 251
Scott, David, 73–74, 84, 98, 100
Sea of Rains (Moon), 72, *73*, 87, *92*, 93, 98
Sea of Serenity (Moon), 77–79, 87, 105, *Plate VII*
Sea of Tranquillity (Moon), 68, 78, 93, *Plate VII*
seamount, 13, *13*, 221
sector collapse, *see* collapse
Semeru (volcano, Earth), 8
Shepard, Alan, 72, 84
Shergotty, shergottite (meteorite, Mars), 155–58
shield volcano
  on Earth, 48, 51
  on Io, 279, *280*, 297, 323
  on Mars, *131*, 132, 134, 137, 150, 169, *174–75*, 175–81, *184*, 184–86, *190*, 190–91, *192*, 197, 199, *Plates XII, XVI*
  on the Moon, 95–96, *95*, *111–12*, 113, 124
  on Venus, 212, 215, 217, 219, *222–23*, 222–24, 227, 229, *233*, *240*, 244–45, 267, *Plates XVIII, XX*
Shombole (volcano, Earth), 57
Shorty crater (Moon), 79, *79*–82, 108–110, *109*, *Plates IX, X*
Siberian Traps, 28–30, 136
Sif Mons (volcano, Venus), *208*, 223, *240*, 255, *Plate XX*

silica, siliceous, 17–18, 36, 141, 144, 148, 212, 229, 254, 261, 294
sill, 224
sinuous rille, 44, 67, 72–76, 73, 75, 94, 111, 112, 113–15, 121–25, 121, 123–25
Siretoko-Iosan (volcano, Earth), 283
Snake River plain (Earth), 96, 229
SNC (meteorites, Mars), 154–58, 155
sodium, 55, 67, 69, 152, 284
soil
    of Mars, see Mars, soil
    of the Moon, see regolith
Sojourner, 148, 149
Somerville (impact crater, Venus), 246–47, 246
Soufrière, La (volcano, Earth), 118
South Pole-Aitken basin, Moon, 87
spatter cone, 44, 54, 56, 57–58, 113–14
Spirit, 151–54
Stealth region (Mars), 185
stellate fracture center, 224–25
stratovolcano, 55, 57
Stromboli (volcano, Earth), 8, 22
strombolian (eruption), 22, 53, 55, 57, 232, 234
subalkaline lava, 212–13
subduction
    on Earth, 14–15, 22, 59, 141, 148
    on Mars, 167
    on Venus, 218–19
sulfate
    on Earth, 163, 283
    on Mars, 145, 152
sulfide
    on the Earth, 163
    on Mars, 164
    on Venus, 258
sulfur
    on Earth, 20, 31, 44, 283
    on Io, 282, 285, 287, 290–91, 298, 307–308, 318, 320–21, 328
    on Mars, 144–45, 149, 152
    on Venus, 240
sulfur dioxide,
    on Earth, 24–25, 29, 44
    on Io, 274, 280, 284–85, 288, 289–90, 293, 297–98, 301, 303, 307–308, 316, 318, 320, 321, 328, Plate XXI
    on Venus, 213, 239, 258
sulfur flow,
    on Earth, 282–83

on Io, 282–83, 323–25, 327, 330
sulfuric acid
    on Mars, 152
    on Venus, 209, 214, 240
superoxide (on Mars), 144
Surt (volcano, Io), 278, 281, 285, 289, 304
Surveyor, 67
Surveyor 3, 70, 71, 244
Surveyor 7, 68
Syrtis Major (Mars), 150

Tambora (volcano, Earth), 5, 24–25, 27, 27
Taurus-Littrow (valley, Moon), 76–78, 77, 94, 104–111, 104, 106–107, Plates IX, X
tectonic plates, see plate tectonics
Telemann Formation (Moon), 122
temperature (effect on eruption), 209, 230–32
terra(e) (lunar highland), 64
Terra Cimmeria (Mars), 166
Terra Meridiani (Mars), 151–52
Terra Sirenum (Mars), 166
tessera (Venus), 236, 238, 240, 261
Tharsis (plateau, Mars), 131–34, 135, 141, 158, 168, 236, Plate XI
Tharsis Tholus (volcano, Mars), 128, 134, 135, 147
Theia Mons (volcano, Venus), 208, 215, 221, 223, 242–47, 242–43, 246
Themis Regio (Venus), 236
Theodora Patera (volcano, Venus), 222
tholeiitic basalt, 36, 41, 45, 50, 211, 212–13, 230
tholus (pl.: tholi) (volcanic construct)
    on Mars, 134, 197
thorium, 7, 210
tick (Venus), 228, 229, 254
tidal stress, 64, 87, 275–76, 292–93
titanium, 69, 71, 81, 83–84, 108, 118, 144, Plate VII
titanium basalt, 70, 84, 87, 98, 105, 113, 122
Toba (volcano, Earth), 8, 19, 26
trachyandesite, 18, 52
trachybasalt, 48, 51, 53
trachyte, 18, 41, 52, 210, 264
tranquillityite, 84
transient phenomenon (on Moon), 66, 93, 121–22
trap, 28–30
tuff (cone, ring), 23, 37

tube, *see* lava tube
tunnel, *see* lava tube
Tupan (volcano, Io), *271*, 303–304, 319,
    Plate XXIV
tuya, 39, 171
Tvashtar (volcano, Io), *271*, 297–304, *299*,
    302, 304
Tycho (crater, Moon), *68*, 108
Tyrrhena Patera (volcano, Mars), *128*, 138,
    139,

ultramafic (lava), 290–92, 316
Ulysses Patera (volcano, Mars), *128*, 134,
    135, *146*
Utopia (plains, Mars), *143*, 144
uranium, 7, 70, 122, 210
Uranius Patera (volcano, Mars), *128*, 134,
    135
Uranius Tholus (volcano, Mars), *128*, 134,
    135, *147*
Uwekakuna laccolith, 44

Valle del Bove (Etna, Earth), 48, *49*, 52, 124,
    247
Valles (caldera, Earth), 19
Valles Marineris (Mars), 150, *151*, 168–71
valley network (Mars), 167–68, 194
Vega 1 and *2*, 213
Venera, 210–213
Venera 1–4, 210
Venera 7, 210
Venera 8, 210
Venera 9, 210, 212, 244–45, *245*
Venera 10, 210, 212
Venera 11 and *12*, 212
Venera 13, *211*, 212–13, *Plate XIX*
Venera 14, *211*, 212–13
Venera 15 and *16*, 215–17, 219
vent (eruptive), *37*, 72, 76–77, *88*, 142, 171,
    186, *199*, 262, 277, 285, 297, 300,
    302–303, 307, *310*, 318, 320–21
Venus
    age of volcanism on, 236–39, 244, 250,
        257, 266
    arachnoids, 217, 219, 224–25, *226*
    ash falls, *233*, 233–34, 250, 264
    atmosphere, 207, 209–10, 213–14, *214*,
        230, 245
    calderas, 215–17, *216*, 222, *222*, *240*,
        244, 247, 250–52, 256–57, 267
    channels, 217, 219, 229–31, *231*, 265–69

clouds, 209, 245
cones, 219, 221, 250
core, 239
coronae, 216–17, 219, 221, 224–27, *224*,
    225–27, 255–60
crust, 219, 232, 238–241
domes, 217, 224, *224*, 227–29, *228*, *235*,
    250, 252, 254, 260–64, *260–61*,
    263–64
erosion, 209
eruptions, 209, 213–14, 232–34, 238–39
greenhouse effect, 207, 209, 239–41
hot spots, 209, 213, 219, 225, 235–36,
    244, 249, 256–57
impact craters, 237–38, *237*, 251, 257
lava flows, 210–13, *211*, 215–17,
    219–21, *216*, 227, 229–30, 232, *235*,
    237, 244–47, 249–51, *252–54*,
    254–57, *255–256*, 259, 265–269,
    265–66, *270*
mantle, 239–40
map of, *208*
orbital parameters, 207
plate tectonics, 20, 209, 217–19, 235,
    239
plumes (mantle), 219, 221, 225, 235,
    249, 259–60, 262
rift zones, 210–13, *211*, 215–17, 219–21,
    216, 227, 229–30, 232, *235*, *237*,
    244–47, 249–51, *252–54*, 254–57,
    255–256, 259, *Plate XVII*
shields, 212, 215, 217, 219, *222–23*,
    222–24, 227, 229, *233*, *240*, 244–45,
    267, *Plates XVII, XX*
topography, *Plate XVII*
volcano classification, *222*, 222–24
volcano population, 215–17, 219–27,
    234–36
vesicle (in lava), 69, *82*, 106, 143, *143*, 148
Vesuvius, Mount (volcano, Earth), 22, 93,
    213
Viking, 143–44, 156, 161
Virunga (volcanoes, Earth), *14*
viscosity (of lava), 21–22, 66, 94, 269, 284
Viti (maar, Earth), *35*, 38
volatiles (in magma), 18–27, 69–70, 105,
    115, 138, 140–41, 168, 170, 203,
    232–34, 250, 254
volcanic winter, 28
volcano, *see* cone, dome, shield, and
    individual names

volcano field, 219–21
Volund (volcano, Io), 274, 288
Voyager, 270–81, 284–85, 288–89, 292, 305, 313–14, 318, 323–24, 326
vulcanian (eruption), 22, *23, 55, 58, 234
Vulcano (volcano, Earth), 22, *23*, 263, 283, *283*

water
    in eruption plumes, 24, 165, 168, 170, 239–40
    in lava flows, 24, 297
    in magma, 18, 20, 22, 60, 168, 233
    sapping action, 194
West Rota (seamount, Earth), *13*

wind (on Venus), 210
Worden, Al, 77, 85
wrinkle ridge, 67, *92, 111–12*, 114, 182, 195–96

X-ray spectrometry
    on Mars, 144–45, 148
    on Venus, 212–13

Yellowstone, 163, 195, 279
Yogi (rock, Mars), 148–49
Young, John, 76, 95

Zagami (meteorite, Mars), 158
Zamama (lava field, Io), 289, 297

CPSIA information can be obtained at www.ICGtesting.com
Printed in the USA
BVOW071221181012

303333BV00001B/69/P